Plantwide
Process
Control

Plantwide
Process
Control

William L. Luyben
Department of Chemical Engineering
Lehigh University

Björn D. Tyréus

Michael L. Luyben
Central Research & Development
E. I. du Pont de Nemours & Co., Inc.

McGraw-Hill

New York San Francisco Washington, D.C. Auckland Bogotá
Caracas Lisbon London Madrid Mexico City Milan
Montreal New Delhi San Juan Singapore
Sydney Tokyo Toronto

Library of Congress Cataloging-in-Publication Data

Luyben, William L.
    Plantwide process control / William L. Luyben, Björn D. Tyréus,
Michael L. Luyben
        p.   cm.
    Includes bibliographical references and index.
    ISBN 0-07-006779-1 (acid-free paper)
    1. Chemical process control. I. Tyréus, Björn D. II. Luyben,
Michael L., (date). III. Title.
TP155.75.L894   1998
660'.2518—dc21
                                                    98-16167
                                                        CIP

McGraw-Hill

A Division of The McGraw·Hill Companies

1 2 3 4 5 6 7 8 9 0   FGR/FGR   9 0 3 2 1 0 9 8

ISBN 0-07-006 779-1

The sponsoring editor for this book was Bob Esposito, the editing supervisor was Peggy Lamb, and the production supervisor was Pamela A. Pelton. It was set in Century Schoolbook by ATLIS Graphics.

Printed and bound by Quebecor Fairfield.

This book is printed on recycled, acid-free paper containing a minimum of 50% recycled, de-inked fiber.

McGraw-Hill books are available at special quantity discounts to use as premiums and sales promotions, or for use in corporate training programs. For more information, please write to the Director of Special Sales, McGraw-Hill, 11 West 19th Street, New York, NY 10011. Or contact your local bookstore.

To my parents Anna-Stina and Jean, and to my wife Kerstin and our children Daniel and Martina for inspiring and encouraging.

BDT

To the loving memory of Beatrice D. Luyben.

MLL and WLL

Contents

Preface

The goal of this book is to help chemical engineering students and practicing engineers develop effective control structures for chemical and petroleum plants. Our focus is on the *entire* plant, not just the individual unit operations. An apparently appropriate control scheme for a single reactor or distillation column may actually lead to an inoperable plant when that reactor or column is connected to other unit operations in a process with recycle streams and energy integration.

Our objective is to design a control system that provides basic regulatory control of the process; i.e., the plant will sit where we want it despite disturbances. Above this regulatory structure we can then build systems to improve plant performance: real-time on-line operations optimization (RTO), planning and scheduling, and expert systems, among others. But if the basic regulatory control does not work as the foundation of plant operation, none of the higher level objectives can be met.

Because of the problem's complexity, our approach is heuristic and experiential. The collected years of experience of the authors is rapidly approaching eight decades, so we have been around long enough to have had our tails caught in the wringer many times. But we have learned from the mistakes that we and others have made. The authors have had the good fortune to learn the basics of plantwide control from the grandfather of the technology, Page Buckley of DuPont. Page was a true pioneer in chemical engineering process control. We also have learned from the experience and inventiveness of many practicing control engineers: Greg Shinskey, John Rijnsdorp, Jim Downs, Jim Douglas, Vince Grassi, Terry Tolliver, and Ed Longwell, among others. These individuals have helped in the evolution of concepts and strategies for doing plantwide control.

Although the methods discussed are heuristic, we certainly recommend the use of algorithmic and mathematical techniques where this approach can aid the analysis of the problem. Methods such as singular value decomposition, condition number analysis, and multivariable Nyquist plots have their place in plantwide control. But the primary mathematical tool employed in this book is a rigorous, nonlinear mathematical model of the entire plant. This model must faithfully capture the nonlinearity and the constraints encountered in the plant under consideration. Any plantwide control scheme must be tested on this type of model because linear, unconstrained models are not adequate to predict many of the important plantwide phenomena. So mathematical modeling and simulation are vital tools in the solution of the plantwide control problem.

Fortunately we now stand at the dawn of a new era in which the computer-aided engineering software tools and computer horsepower permit engineers to assemble a flowsheet, perform the steady-state analysis (mass and energy balances, engineering economics, and optimization), and then evaluate the dynamic performance of the plant. Commercial software packages that combine steady-state and dynamic models represent a major breakthrough in the tools available to the process engineer and to the control engineer. Actually we predict that in not too many years these two functions will be combined and will be performed (as they should be) by the same individual. An appreciation of dynamics is vital in steady-state design and an appreciation of design is vital in process control.

Four detailed case studies of realistically complex industrial-scale processes are discussed in this book. Models of three of these have been developed by Aspen Technology and Hyprotech in their commercial simulators and are available directly from the vendors. These models may be obtained electronically from the Web sites: www.aspentec.com and www.hyprotech.com. We appreciate the efforts expended by these companies in making these case studies available to students and engineers. The methods developed in this book are independent of the simulation software used to model the plant.

The concepts presented in this book can be applied at all levels of control engineering: in the conceptual development of a new process, in the design of a grass-roots commercial facility, in debottlenecking and plant revamps, and in the operation of an existing process. However, the emphasis is on new plant design because this is the level at which the effect of considering plantwide control can have the most significant impact on business profitability. The cost of modifying the process at the design stage is usually fairly low and the effect of these modifications on the dynamic controllability can be enormous. Old war stories abound in the chemical industry of plants that have never run because of dynamic operability problems not seen in a steady-state flowsheet, with millions of dollars going down the drain.

This book is intended for use by students in senior design courses in which dynamics and control are incorporated with the traditional steady-state coverage of flowsheet synthesis, engineering economics, and optimization. A modern chemical engineering design course should include all three aspects of design (steady-state synthesis, optimization, and control) if our students are going to be well-prepared for what they will deal with in industry.

This book also should be useful to practicing engineers, both process engineers and control engineers. Most engineers have had a control course in their undergraduate and/or graduate training. But many of these courses emphasize the mathematics of the subject, giving very little if any coverage of the important practical aspects of designing effective control structures. Most of the control textbooks have very limited treatments of control system design, even for individual units. There are no textbooks that cover the subject of plantwide process control in a quantitative practical way. We strive to fill that gap in technology with this book.

We hope you find the material interesting, understandable, and useful. We have developed and applied the methods discussed in this book for many years on many real industrial processes. They work!

But don't expect this book to free you of the need to think! We do not provide a black box into which you simply feed the input data and out comes a "globally optimum" solution. The problem here is an open-ended *design problem* for which there is no single "correct" answer. Our procedure requires the application of thought, insight, process understanding, and above all, practice on realistic problems such as those provided in this book. These ingredients should lead you to an effective control structure. There is no claim that this control structure is necessarily the best. But it should provide stable regulatory control of the plant.

Thanks are due to a number of individuals who have contributed to the development of the technology outlined in this book. The legacy of Page Buckley is apparent on almost every page. Lehigh students, both undergraduate and graduate, have contributed significantly to the development of this book by their youthful enthusiasm, willingness to work hard, and interest in real engineering problems. They have provided the senior author with enough job satisfaction to offset the frustrations of dealing with university bureaucrats.

In addition to the legacy of Lehigh University (as well as Princeton University and Prof. C. A. Floudas), B. D. Tyréus and M. L. Luyben want to acknowledge DuPont and its culture of technological innovation and excellence. We have had the opportunity to work on and learn about many different processes, and we have tried in this book to synthesize in some coordinated way part of our experiences. Most of this book is inspired by our work over the years with many outstanding process and control engineers at DuPont, who have taught us so much. Listing them all would require considerable space and leave us vulnerable to overlooking someone. Nonetheless, they know who they are and we thank each of them. We also could not have written this book without the leadership provided by James A. Trainham, Roger A. Smith, and W. David Smith, Jr.

William L. Luyben
Björn D. Tyréus
Michael L. Luyben

Basics

1

Introduction

1.1 Overview

Plantwide process control involves the systems and strategies required to control an entire chemical plant consisting of many interconnected unit operations.

One of the most common, important, and challenging control tasks confronting chemical engineers is: How do we design the control loops and systems needed to run our process? We typically are presented with a complicated process flowsheet containing several recycle streams, energy integration, and many different unit operations: distillation columns, reactors of all types, heat exchangers, centrifuges, dryers, crystallizers, liquid-liquid extractors, pumps, compressors, tanks, absorbers, decanters, etc. Given a complex, integrated process and a diverse assortment of equipment, we must devise the necessary logic, instrumentation, and strategies to operate the plant safely and achieve its design objectives.

This is, in essence, the realm of control system synthesis for an entire plant. What issues do we need to consider? What is of essential importance within this immense amount of detail? How does the dynamic behavior of the interconnected plant differ from that of the individual unit operations? What, if anything, do we need to model or test? How do we even begin?

This book addresses each of these questions and explains the fundamental ideas of control system synthesis. As its core, the book presents a general heuristic design procedure that generates an effective plantwide base-level regulatory control structure for an *entire, complex* process flowsheet and not simply individual units.

The nine steps of the design procedure center around the fundamental principles of plantwide control: energy management; production rate; product quality; operational, environmental, and safety constraints; liquid level and gas pressure inventories; makeup of reactants; component balances; and economic or process optimization.

We first review in Part 1 the basics of plantwide control. We illustrate its importance by highlighting the unique characteristics that arise when operating and controlling complex integrated processes. The steps of our design procedure are described. In Part 2, we examine how the control of individual unit operations fits within the context of a plantwide perspective. Reactors, heat exchangers, distillation columns, and other unit operations are discussed. Then, the application of the procedure is illustrated in Part 3 with four industrial process examples: the Eastman plantwide control process, the butane isomerization process, the HDA process, and the vinyl acetate monomer process.

1.2   HDA Process

Let's begin with an example of a real industrial process to highlight what we mean by *plantwide process control*. The hydrodealkylation of toluene (HDA) process is used extensively in the book by Douglas (1988) on conceptual design, which presents a hierarchical procedure for generating steady-state flowsheet structures. Hence the HDA process should be familiar to many chemical engineering students who have had a course in process design. It also represents a flowsheet topology that is similar to many chemical plants, so practicing engineers should recognize its essential features.

The HDA process (Fig. 1.1) contains nine basic unit operations: reactor, furnace, vapor-liquid separator, recycle compressor, two heat exchangers, and three distillation columns. Two vapor-phase reactions are considered to generate benzene, methane, and diphenyl from reactants toluene and hydrogen.

$$\text{Toluene} + \text{H}_2 \rightarrow \text{benzene} + \text{CH}_4 \qquad (1.1)$$

$$2\text{Benzene} \rightleftharpoons \text{diphenyl} + \text{H}_2 \qquad (1.2)$$

The kinetic rate expressions are functions of the partial pressures of toluene p_T, hydrogen p_H, benzene p_B, and diphenyl p_D, with an Arrhenius temperature dependence. By-product diphenyl is produced in an equilibrium reaction.

$$r_1 = k_{1(T)} p_T p_H^{1/2} \qquad (1.3)$$

$$r_2 = k_{2f(T)} p_B^2 - k_{2r(T)} p_D p_H \qquad (1.4)$$

The two fresh reactant makeup feed streams (one gas for hydrogen

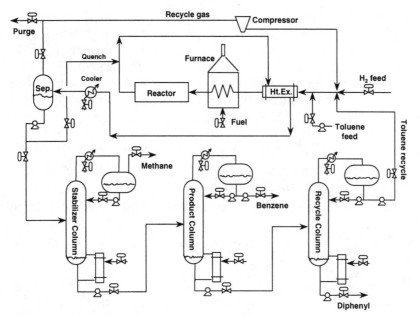

Figure 1.1 HDA process flowsheet.

and one liquid for toluene) are combined with the gas and liquid recycle streams. This combined stream is the cold inlet feed to the process-to-process heat exchanger, where the hot stream is the reactor effluent after the quench. The cold outlet stream is heated further, via combustion of fuel in the furnace, up to the required reactor inlet temperature. The reactor is adiabatic and must be run with an excess of hydrogen to prevent coking. The reactor effluent is quenched with liquid from the separator to prevent fouling in the process-to-process heat exchanger.

The hot outlet stream from the process-to-process heat exchanger goes to a partial condenser and then to a vapor-liquid separator. The gas stream from the overhead of the separator recycles unconverted hydrogen plus methane back to the reactor via a compressor. Since methane enters as an impurity in the hydrogen feed stream and is further produced in the reactor, it will accumulate in the gas recycle loop. Hence a purge stream is required to remove methane from the process. Part of the liquid from the separator serves as the reactor quench stream.

The remainder of the liquid from the separator is fed to the stabilizer column to remove any of the remaining hydrogen and methane gas from the aromatic liquids. The bottoms stream from the stabilizer column feeds the product column, which yields the desired product benzene in the distillate. The by-product diphenyl exits from the process in the bottoms stream from the recycle column, which is fed from the bottoms of the product column. The liquid distillate stream from the recycle column returns unconverted toluene to the reactor.

Given this process flowsheet, we'd like to know how we can run this process to make benzene. We naturally have a lot of questions we want answered about operating this plant:

- How do we control the reactor temperature to prevent a runaway?
- How can we increase or decrease the production rate of benzene depending upon market conditions?
- How do we ensure the benzene product is sufficiently pure for us to sell?
- How do we know how much of the fresh hydrogen and toluene feed streams to add?
- How do we determine the flowrate of the gas purge stream?
- How can we minimize the raw material yield loss to diphenyl?
- How do we prevent overfilling any liquid vessels and overpressuring any units?
- How do we deal with units tied together with heat integration?
- How can we even test any control strategy that we might develop?

Answering these questions is not at all a trivial matter. But these issues lie at the foundation of control system synthesis for an entire plant. The plantwide control problem is extremely complex and very much open-ended. There are a combinatorial number of possible choices and alternative strategies. And there is no unique "correct" solution.

Reaching a solution to the complex plantwide control problem is a creative challenge. It demands insight into and understanding of the chemistry, physics, and economics of real processes. However, it is possible to employ a systematic strategy (or engineering method) to get a feasible solution. Our framework in tackling a problem of this complexity is based upon heuristics that account for the unique features and concerns of integrated plants. This book presents such a general plantwide control design procedure.

The scope embraces continuous processes with reaction and separation sections. Because our approach in this book is based upon a plantwide perspective, we cover what is relevant to this particular area. We omit much basic process control material that constitutes the framework and provides the tools for dynamic analysis, stability, system identification, and controller tuning. But we refer the interested reader to Luyben and Luyben (1997) and other chemical engineering textbooks on process control.

1.3 History

Control analysis and control system design for chemical and petroleum processes have traditionally followed the "unit operations approach" (Stephanopoulos, 1983). First, all of the control loops were established individually for each unit or piece of equipment in the plant. Then the pieces were combined together into an entire plant. This meant that any conflicts among the control loops somehow had to be reconciled. The implicit assumption of this approach was that the sum of the *individual* parts could effectively comprise the *whole* of the plant's control system. Over the last few decades, process control researchers and practitioners have developed effective control schemes for many of the traditional chemical unit operations. And for processes where these unit operations are arranged in series, each downstream unit simply sees disturbances from its upstream neighbor.

Most industrial processes contain a complex flowsheet with several recycle streams, energy integration, and many different unit operations. Essentially, the plantwide control problem is how to develop the control loops needed to operate an *entire* process and achieve its design objectives. Recycle streams and energy integration introduce a feedback of material and energy among units upstream and downstream. They also interconnect separate unit operations and create a path for disturbance propagation. The presence of recycle streams profoundly alters the dynamic behavior of the plant by introducing an integrating effect that is not localized to an isolated part of the process.

Despite this process complexity, the unit operations approach to control system design has worked reasonably well. In the past, plants with recycle streams contained many surge tanks to buffer disturbances, to minimize interaction, and to isolate units in the sequence of material flow. This allowed each unit to be controlled individually. Prior to the 1970s, low energy costs meant little economic incentive for energy integration. However, there is growing pressure to reduce capital investment, working capital, and operating cost and to respond to safety and environmental concerns. This has prompted design engineers to start eliminating many surge tanks, increasing recycle streams, and introducing heat integration for both existing and new plants. Often this is done without a complete understanding of their effects on plant operability.

So economic forces within the chemical industry are compelling improved capital productivity. Requirements for on-aim product quality control grow increasingly tighter. More energy integration occurs. Improved product yields, which reduce raw material costs, are achieved via lower reactant per-pass conversion and higher material recycle rates through the process. Better product quality, energy integration, and higher yields are all economically attractive in the steady-state flowsheet, but they present significant challenges to smooth dynamic plant operation. Hence an effective control system regulating the entire plant operation and a process designed with good dynamic performance play critical parts in achieving the business objectives of reducing operating and capital costs.

Buckley (1964) proposed a control design procedure for the plantwide control problem that consisted of two stages. The first stage determined the material balance control structure to handle vessel inventories for low-frequency disturbances. The second established the product quality control structure to regulate high-frequency disturbances. This procedure has been widely and effectively utilized. It has served as the conceptual framework in many subsequent ideas for developing control systems for complete plants. However, the two-stage Buckley procedure provides little guidance concerning three important aspects of a plantwide control strategy. First, it does not explicitly discuss energy management. Second, it does not address the specific issues of recycle systems. Third, it does not deal with component balances in the context of inventory control. By placing the priority on material balance over product quality controls, the procedure can significantly limit the flexibility in choosing the latter.

We believe that chemical process control must move beyond the sphere of unit operations into the realm of viewing the plant as a whole system. The time is ripe in the chemical and petroleum industry for the development of a plantwide control design procedure. The technology, insight, and understanding have reached a state where general guidelines can be presented. The computer software needed for plantwide dynamic simulations is becoming commercially available. While linear methods are very useful to analyze control concepts, we strongly believe that the final evaluation of any plantwide control structure requires rigorous nonlinear dynamic simulations, not linear transfer function analysis.

1.4   Model-Based and Conventional Control

Some people claim that the plantwide control problem has already been solved by the application of several commercial forms of model predictive control (MPC). MPC rests on the idea that we have a fair amount of knowledge about the dynamic behavior of the process and that this knowledge can be incorporated into the *controller* itself. The controller uses past information and current measurements to predict

the future response and to adjust its control valves so that this antici-
pated response is optimal in some sense.

Model predictive control is particularly useful when several control
valves (or manipulators) affect an output of interest (what is called
interaction) and also when some sort of constraint comes into play
either on the inputs or on some measured variable. Since the controller
itself *knows* about these interactions and constraints, it can in theory
avoid those perils. It is important to remember that MPC merely suggests that the controller can predict the process response into the future,
only to be checked (and corrected) by the next round of measurements.

On the other hand, conventional control approaches also rely on
models, but they are usually not built into the controller itself. Instead
the models form the basis of simulations and other analysis methods
that guide in the selection of control loops and suggest tuning constants
for the relatively simple controllers normally employed [PI, PID, I-only,
P-only, lead-lag compensation, etc. (P = proportional, PI = proportional-
integral, PID = proportional-integral-derivative)]. Conventional control approaches attempt to build the *smarts* into the *system* (the process
and the controllers) rather than only use complex control algorithms.

Our understanding is that MPC has found widespread use in the
petroleum industry. The chemical industry, however, is still dominated
by the use of distributed control systems implementing simple PID
controllers. We are addressing the plantwide control problem within
this context. We are not addressing the application of multivariable
model-based controllers in this book.

Very few unbiased publications have appeared in the literature comparing control effectiveness using MPC versus a well-designed conventional control system. Most of the MPC applications reported have
considered fairly simple processes with a small number of manipulated
variables. There are no published reports that discuss the application
of MPC to an entire complex chemical plant, with one notable exception.
That is the work of Ricker (1996), who compared MPC with conventional
PI control for the Eastman process (TE problem). His conclusion was
"there appears to be little, if any, advantage to the use of nonlinear
model predictive control (NMPC) in this application. In particular, the
decentralized strategy does a better job of handling constraints—an
area in which NMPC is reputed to excel."

One of the basic reasons for his conclusion ties into the plantwide
context that our procedure explicitly addresses, namely the need to
regulate all chemical inventories. MPC gives no guidance on how to
make the critical decisions of what variables need to be controlled. As
Ricker states, "the naive MPC designer might be tempted to control only
variables having defined setpoints, relying on optimization to make
appropriate use of the remaining degrees of freedom. This fails in the

TE problem. As discussed previously, all chemical inventories must be regulated; it cannot be left to chance. Unless setpoints for key internal concentrations are provided, MPC allows reactant partial pressures to drift to unfavorable values." Our design procedure considers the concept of component balances as an explicit step in the design.

Another reason is related to the issue of constraints and priorities, which we address in the sequence of steps for our design procedure. Ricker says that "the TE problem has too many competing goals and special cases to be dealt with in a conventional MPC formulation." Normally this is addressed within MPC by the choice of weights, but for the Eastman process the importance of a variable changes depending upon the situation. "Ricker and Lee found that no single set of weights and constraints could provide the desired performance in all cases."

While we use conventional control systems here, our plantwide control design procedure does not preclude the use of MPC at a certain level. Our focus is on the issues arising from the operation of an integrated process. We find that a good control structure provides effective control, independent of any particular controller algorithm, while a poor one cannot be greatly improved with any algorithm (MPC or PID controllers).

1.5 Process Design

The traditional approach to developing a new process has been to perform the design and control analyses sequentially. First, the design engineer constructs a steady-state process flowsheet, with particular structure, equipment, design parameters, and operating conditions. The objective is to optimize the economics of the project in evaluating the enormous number of alternatives. The hierarchical design procedure proposed by Douglas (1988) is a way to approach this task. Little attention is given to dynamic controllability during the early stages of the design.

After completion of the detailed design, the control engineer then must devise the control strategies to ensure stable dynamic performance and to satisfy the operational requirements. The objective is to operate the plant in the face of potentially known and unknown disturbances, production rate changes, and transitions from one product to another.

While this staged approach has long been recognized as deficient, it is defensible from a certain perspective. For example, it would be difficult for the control engineers to specify the instrumentation and the distributed control system (DCS) without knowing exactly what process it was intended for. Similarly, it would make no sense for the process engineers to request a control system design for all those flowsheets that were considered but rejected on the basis of steady-state economics alone. However, this staged approach can result in missed opportunities because of the close connection between process design and controllability. How a process is designed fundamentally determines its inherent controllability, which means qualitatively how well the process rejects disturbances and how easily it moves from one operating condition to another. In an ideal project system, dynamics and control strategies would be considered during the process synthesis and design activities.

This issue grows increasingly important as plants become more highly integrated with complex configurations, recycle streams, and energy integration. Competitive economic pressures, safety issues, and environmental concerns have all contributed to this. However, if a control engineer becomes involved early enough in the process design, he or she may be able to show that it would be better in the long run to build a process with higher capital and utility costs if that plant provides more stable operation and less variability in the product quality.

We believe that process design impacts controllability far more than control algorithms do. We base our opinion on many years of experience. We have participated as control engineers in many design projects. Some involved building new plants with new process technology, some involved new plants with existing technology, and some projects were modernizations of the control system on an existing plant. We have found that a consideration of dynamics and control strategies for new process designs has a much larger positive economic impact (when the design can potentially be modified) compared with control strategy upgrades on an existing process (with a fixed design). However, we stress that for those new plants and technologies we became involved before the process design was fixed. We performed dynamic simulations and undertook control system design as soon as the process engineers had an economically viable flowsheet. Most importantly, by working together with the process engineers and plant engineers, we changed the flowsheet until we were all satisfied that we had developed the most profitable process when viewed over the entire life time of the project. This inevitably involved making trade-offs between steady-state investment economics and dynamic performance measured in uptime, throughput, product quality, and yield.

One of the important themes weaving through this book is the central role we place on the process design. Good control engineers need also to be good process engineers!

1.6 Spectrum of Process Control

We can view the field of process control as five parts of a continuous spectrum (Fig. 1.2). Each part is important, can be economically signifi-

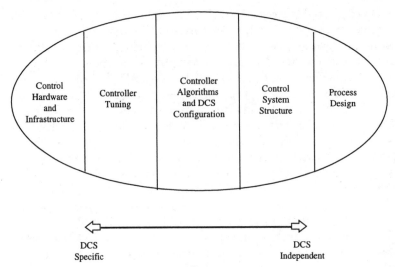

Figure 1.2 Spectrum of process control.

cant, and interacts in some manner with the others. Moving toward the left on the spectrum means dealing with more detailed issues on the level of the distributed control system (DCS). Moving toward the right means operating on a more general level with issues that are independent of the DCS.

The far left part of the spectrum deals with the control hardware and infrastructure required to operate a plant. We need to assemble the proper types of control valves and process measurements (for temperature, flow, pressure, composition, etc.). These are the sensory devices of the plant and are essential for any control system to function. Any control strategy, no matter how clever, will have severe difficulties without the right measurements and valves in the process. An Instrument Society of America (ISA) publication catalog (67 Alexander Drive, P.O. Box 12277, Research Triangle Park, NC 27709) contains many references that deal with control hardware.

The next part involves controller tuning. We must determine the tuning constants for the controllers in the plant. While this task is often performed by using heuristics and experience, it can sometimes be a nontrivial exercise for certain loops. We recommend using a relay-feedback test that determines the ultimate gain and period for the control loop, from which controller settings can be calculated (Luyben and Luyben, 1997).

The middle of the spectrum deals with the controller algorithms and DCS configuration. We must decide the type of controller to use (proportional, integral, derivative, multivariable, nonlinear, model predictive, etc.). We must also determine whether we need dynamic elements (lead/lags, feedforward, etc.) and how to handle overrides and interlocks. In addition, input and output variables must be assigned loop numbers, displays must be created, alarms must be specified, instrument groupings must be determined, etc.

The next part is the determination of the control system structure. We must decide what variables to control and manipulate and how these should be paired. The control structure is vitally important because a poor strategy will result in poor performance no matter what type of control algorithm we use or how much we tune it. There is little information or guidance in the literature or in process control textbooks (both introductory and advanced) on how to develop an effective control structure for an entire complex chemical plant. This is the main subject of this book.

The far right part of the spectrum is the design of the process itself. We sometimes can change the flowsheet structure, use different design parameters, and employ different types of process equipment to produce a plant that can be controlled more easily than other alternatives. At this level, a good process control engineer can potentially have an enormous economic impact. Most companies in the chemical and petroleum industries have had the unfortunate and unwelcome experience of building a plant that could not easily be started up because of operational difficulties arising from the plant design. Fixing these kinds of problems after the plant is built can often require large amounts of additional capital expense in addition to the lost sales opportunities.

In this book, we focus primarily on control structure selection. Interactions between design and control are illustrated by examples, and the effects of design parameters on control are discussed. However, we do not present a synthesis procedure for process design that is capable of generating the most controllable flowsheet for a given chemistry. This is still very much an open area for further research.

1.7 Conclusion

In this first chapter we have defined the plantwide process control problem. This was illustrated by using the HDA process, which will figure prominently in later parts of the book. We have provided a historical perspective and context. Finally we explained where the material in this book fits into the spectrum of process control activities.

1.8 References

Buckley, P. S. *Techniques of Process Control*, New York: Wiley (1964).
Douglas, J. M. *Conceptual Design of Chemical Processes*, New York: McGraw-Hill (1988).

Luyben, W. L., and Luyben M. L. *Essentials of Process Control*, New York: McGraw-Hill (1997).

Ricker, N. L. "Decentralized Control of the Tennessee Eastman Challenge Process," *J. Proc. Cont.*, **6**, 205–221 (1996).

Stephanopoulos, G. "Synthesis of Control Systems for Chemical Plants—A Challenge for Creativity," *Comput. Chem. Eng.*, **7**, 331–365 (1983).

2

Plantwide
Control Fundamentals

2.1 Introduction

In this chapter we examine some of the fundamental features and properties of the plantwide control problem. Our goal is to explain why we must design a control system from the viewpoint of the entire plant and not just combine the control schemes of each individual unit.

A typical chemical plant flowsheet has a mixture of multiple units connected both in series and in parallel. As noted in the previous chapter, the common topology consists of reaction sections and separation sections. Streams of fresh reactants enter the plant by being fed into the reaction section (or sometimes into the separation section) through a heat exchanger network. Here the chemical transformations occur to produce the desired species in one or more of a potentially wide array of reactor types: continuous stirred tank, tubular, packed bed, fluidized bed, sparged, slurry, trickle bed, etc.

The reactor effluent usually contains a mixture of reactants and products. It is fed into a separation section where the products are separated by some means from the reactants. Because of their economic value, reactants are recycled back to upstream units toward the reactor. The products are transported directly to customers, are fed into storage tanks, or are sent to other units for further processing. The separation section uses one or more of the fundamental unit operations: distillation, evaporation, filtration, crystallization, liquid-liquid extraction, adsorption, absorption, pressure-swing adsorption, etc. In this book we typically use distillation as the separation method because of its widespread use and our considerable experience with it. Everyone is a victim of his or her experience. Our backgrounds are in petroleum processing and chemical manufacturing, where distillation, despite frequently occurring predictions to the contrary, remains the premier separation method. However, the general principles also apply to processes with other separation units.

In addition to recycle streams returned back to upstream units, thermal integration is also frequently done. Energy integration can link units together in locations anywhere in the flowsheet where the temperature levels permit heat transfer to occur. The reaction and separation sections are thus often intimately connected. If conditions are altered in the reaction section, the resulting changes in flowrates, compositions, and temperatures affect the separation section and vice versa.

Changes in temperatures and thermal conditions can propagate into the separation section and significantly degrade dynamic performance. Changes in flowrates create load disturbances that can be recycled around a material loop. Changes in stream compositions fed into the separation section are also troublesome disturbances because they alter separation requirements (the work of separation is often a strong function of the feed mixture composition). Significant shifts in the compositions and flowrates within the separation section are needed to achieve the desired purities of product and recycle streams. Achieving a composition change can sometimes take a long time because the component inventories within the separation section must be varied and this inherently governs the system's dynamic behavior.

So we must pay particular attention to the effects of the reaction section on the separation section. In this chapter we strip away all of the confusing factors associated with complex physical properties and phase equilibrium so that we can concentrate on the fundamental effects of flowsheet topology and reaction stoichiometry. Therefore, in the processes studied here, we use such simplifying assumptions as constant relative volatilities, equimolal overflow, and constant densities.

These "ideal" physical property assumptions may appear to represent an overly simplistic view of the problem. Our experience, however, is that we can often gain significant insight into the workings and interactions of processes with recycle streams by not confusing the picture with complexities such as azeotropes. Considering the complexities of a real chemical system is, of course, vital at some stage. But we attempt in this chapter to focus on the "forest" and not on the individual "trees."

For example, suppose there is a stream in the process that is a binary mixture of chemical components A and B. If these components obey ideal vapor-liquid equilibrium behavior, we can use a single distillation column to separate them. If they form an azeotrope, we may have to use a two-column separation scheme. If the azeotropic composition changes significantly with pressure, we can use a two-column sequence with each column operating at different pressures. If the azeotrope is homogeneous and minimum boiling, the two fairly pure product streams can be produced as bottoms products from the two columns. So there are two columns in the nonideal case instead of one column in the ideal case. But the reaction section and the recycle streams really don't care if we have one column or two. The reactor sees the same types of disturbances coming from the separation section, perhaps with different dynamics but with similar steady-state effects. Since many of the important plantwide and recycle effects are really steady-state phenomena, the idealized single-column separation section yields results that are similar to those of the complex two-column separation section.

2.2 Integrated Processes

Three basic features of integrated chemical processes lie at the root of our need to consider the entire plant's control system: (1) the effect of material recycle, (2) the effect of energy integration, and (3) the need to account for chemical component inventories. If we did not have to worry about these issues, then we would not have to deal with a complex plantwide control problem. However, there are fundamental reasons why each of these exists in virtually all real processes.

2.2.1 Material recycle

Material is recycled for six basic and important reasons.

1. *Increase conversion:* For chemical processes involving reversible reactions, conversion of reactants to products is limited by thermodynamic equilibrium constraints. Therefore the reactor effluent by necessity contains both reactants and products. Separation and recycle of reactants are essential if the process is to be economically viable.

2. *Improve economics:* In most systems it is simply cheaper to build a reactor with incomplete conversion and recycle reactants than it is to reach the necessary conversion level in one reactor or several in series. The simple little process discussed in Sec. 2.6 illustrates this for a binary system with one reaction $A \to B$. A reactor followed by a stripping column with recycle is cheaper than one large reactor or three reactors in series.

3. *Improve yields:* In reaction systems such as $A \to B \to C$, where B is the desired product, the per-pass conversion of A must be kept low to avoid producing too much of the undesirable product C. Therefore

the concentration of B is kept fairly low in the reactor and a large recycle of A is required.

4. *Provide thermal sink:* In adiabatic reactors and in reactors where cooling is difficult and exothermic heat effects are large, it is often necessary to feed excess material to the reactor (an excess of one reactant or a product) so that the reactor temperature increase will not be too large. High temperature can potentially create several unpleasant events: it can lead to thermal runaways, it can deactivate catalysts, it can cause undesirable side reactions, it can cause mechanical failure of equipment, etc. So the heat of reaction is absorbed by the sensible heat required to raise the temperature of the excess material in the stream flowing through the reactor.

5. *Prevent side reactions:* A large excess of one of the reactants is often used so that the concentration of the other reactant is kept low. If this limiting reactant is not kept in low concentration, it could react to produce undesirable products. Therefore the reactant that is in excess must be separated from the product components in the reactor effluent stream and recycled back to the reactor.

6. *Control properties:* In many polymerization reactors, conversion of monomer is limited to achieve the desired polymer properties. These include average molecular weight, molecular weight distribution, degree of branching, particle size, etc. Another reason for limiting conversion to polymer is to control the increase in viscosity that is typical of polymer solutions. This facilitates reactor agitation and heat removal and allows the material to be further processed.

2.2.2 Energy integration

The fundamental reason for the use of energy integration is to improve the thermodynamic efficiency of the process. This translates into a reduction in utility cost. For energy-intensive processes, the savings can be quite significant. We can illustrate the use and benefits of energy-integration by considering again the HDA process introduced in the previous chapter (Fig. 1.1). Here energy is required to heat up the reactants in the furnace and to provide boilup in the three distillation columns. Heat must be removed in the separator condenser and in the three column condensers. Heat is generated in the exothermic reactor that normally would be removed through the plant utility system. However, by using a feed/effluent heat exchanger we can recover some of that energy. This reduces the amount of fuel required in the furnace to heat up the reactants and the duty required to cool the reactor effluent stream.

In fact we could theoretically introduce considerably more energy

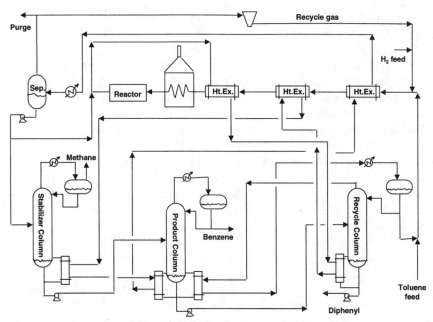

Figure 2.1   HDA process flowsheet with complex heat integration.

integration into the HDA process (Fig. 2.1). This is alternative 6 from the paper by Terrill and Douglas (1987). Heat from the reactor is used in reboilers of all three distillation columns. In addition, condensation of the overhead vapor from the recycle column provides heat input to the base of the product column. This is a good illustration of how units anywhere in the process can be linked together thermally. Figure 2.1 also shows how complex heat-integrated processes can quickly become, creating nontrivial control issues. This highlights why we cannot combine the control systems of individual unit operations in such processes.

2.2.3 Chemical component inventories

We can characterize a plant's chemical species into three types: reactants, products, and inerts. A material balance for each of these components must be satisfied. This is typically not a problem for products and inerts. However, the real problem usually arises when we consider reactants (because of recycle) and account for their inventories within the entire process. Every molecule of reactants fed into the plant must either be consumed via reaction or leave as an impurity or purge. Because of their value, we want to minimize the loss of reactants exiting the process since this represents a yield penalty. So we prevent reactants from leaving. This means we must ensure that every mole of reactant fed to the process is consumed by the reactions.

This is an important concept and is generic to many chemical processes. From the viewpoint of individual units, chemical component balancing is not a problem because exit streams from the unit automatically adjust their flows and compositions. However, when we connect units together with recycle streams, the entire system behaves almost like a pure integrator in terms of the reactants. If additional reactant is fed into the system without changing reactor conditions to consume the reactant, this component will build up gradually within the plant because it has no place to leave the system.

Plants are not necessarily self-regulating in terms of reactants. We might expect that the reaction rate will increase as reactant composition increases. However, in systems with several reactants (e.g., $A + B \rightarrow$ products), increasing one reactant composition will decrease the other reactant composition with an uncertain net effect on reaction rate. Section 2.7 contains a more complete discussion of this phenomenon. Eventually the process will shut down when manipulated variable constraints are encountered in the separation section. Returning again to the HDA process, the recycle column can easily handle changes in the amount of (reactant) toluene inventory within the column. However, unless we can somehow account for the toluene inventory within the entire process, we could feed more fresh toluene into the process than is consumed in the reactor and eventually fill up the system with toluene.

The three features outlined in this section have profound implications for a plant's control strategy. Simple examples in this chapter will illustrate the effects of material recycle and component balancing. Chapter 5 contains more details of the effects created by energy integration on the entire plant.

2.3 Units in Series

If process units are arranged in a purely series configuration, where the products of each unit feed downstream units and there is no recycle of material or energy, the plantwide control problem is greatly simplified. We do not have to worry about the issues discussed in the previous section and we can simply configure the control scheme on each individual unit operation to handle load disturbances.

If production rate is set at the front end of the process, each unit will only see load disturbances coming from its upstream neighbor. If the plant is set up for "on-demand" production, changes in throughput will propagate back through the process. So any individual unit will see load disturbances coming from both its downstream neighbor (flowrate changes to achieve different throughputs) and its upstream neighbor

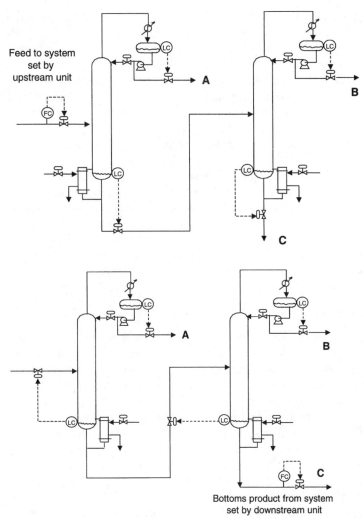

Figure 2.2 Units in series. (*a*) Level control in direction of flow; (*b*) level control in direction opposite flow.

(composition changes as the upstream units adjust to the load changes they see).

Figure 2.2 compares these two possible configurations for a simple plant. A fresh feed stream containing a mixture of chemical components *A*, *B*, and *C* is fed into a two-column distillation train. The relative volatilities are $\alpha_A > \alpha_B > \alpha_C$, and we select the "direct" (or "light-out-first") separation sequence: *A* is taken out the top of the first column and *B* out the top of the second column.

Figure 2.2*a* shows the situation where the fresh feed stream is flow-controlled into the process. The inventory loops (liquid levels) in each unit are controlled by manipulating flows leaving that unit. All disturbances propagate from unit to unit down the series configuration. The only disturbances that each unit sees are changes in its feed conditions.

Figure 2.2*b* shows the on-demand situation where the flowrate of product *C* leaving the bottom of the second column is set by the requirements of a downstream unit. Now some of the inventory loops (the base of both columns) are controlled by manipulating the feed into each column.

When the units are arranged in series with no recycles, the plantwide control problem can be effectively broken up into the control of each individual unit operation. There is no recycle effect, no coupling, and no feedback of material from downstream to upstream units. The plant's dynamic behavior is governed by the individual unit operations and the only path for disturbance propagation is linear along the process.

2.4   Effects of Recycle

Most real processes contain recycle streams. In this case the plantwide control problem becomes much more complex and its solution is not intuitively obvious. The presence of recycle streams profoundly alters the plant's dynamic and steady-state behavior. To gain an understanding of these effects, we look at some very simple recycle systems. The insight we obtain from these idealized, simplistic systems can be extended to the complex flowsheets of typical chemical processes. First we must lay the groundwork and have some feel for the complexities and phenomena that recycle streams produce in a plant.

In this section we explore two basic effects of recycle: (1) Recycle has an impact on the dynamics of the process. The overall time constant can be much different than the sum of the time constants of the individual units. (2) Recycle leads to the "snowball" effect. This has two manifestations, one steady state and one dynamic. A small change in throughput or feed composition can lead to a large change in steady-state recycle stream flowrates. These disturbances can lead to even larger dynamic changes in flows, which propagate around the recycle loop. Both effects have implications for the inventory control of components.

2.4.1   Time constants in recycle systems

Figure 2.3 gives a block-diagram representation of a simple process with recycle. The input to the system is *u*. We can think of this input as a flowrate. It enters a unit in the forward path that has a transfer

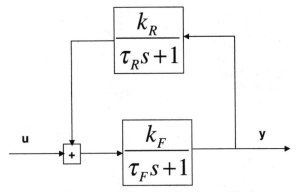

Figure 2.3 Simple block diagram of process with recycle.

function $G_{F(s)}$ that relates dynamically the input to the output of the unit. This transfer function consists of a steady-state gain K_F and a first-order lag with a time constant τ_F:

$$G_{F(s)} = \frac{K_F}{\tau_F s + 1} \tag{2.1}$$

The output of $G_{F(s)}$ is y, which also recycles back through a second transfer function $G_{R(s)}$ in the recycle path. This recycle transfer function also consists of a steady-state gain and a time constant.

$$G_{R(s)} = \frac{K_R}{\tau_R s + 1} \tag{2.2}$$

The output of the recycle block is added to the original input to the process u, and the sum of these two signals enters the forward block $G_{F(s)}$. It is important to note that the recycle loop in this process features *positive* feedback, not *negative* feedback that we are used to dealing with in feedback control. Most recycles produce this positive feedback behavior, which means that an increase in the recycle flowrate causes an increase in the flowrates through the process.

Some simple algebra gives the overall relationship for this system between input and output.

$$\frac{y_{(s)}}{u_{(s)}} = \frac{\dfrac{K_F}{\tau_F s + 1}}{1 - \left(\dfrac{K_F}{\tau_F s + 1}\right)\left(\dfrac{K_R}{\tau_R s + 1}\right)}$$

$$\tag{2.3}$$

$$= \frac{K_F(\tau_R s + 1)}{\tau_F \tau_R s^2 + (\tau_F + \tau_R)s + (1 - K_F K_R)}$$

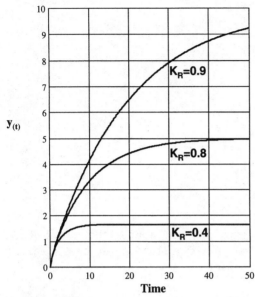

Figure 2.4 Effect of recycle loop gain on overall dynamic response.

The denominator of the transfer function is the characteristic equation of any system, so the characteristic equation of this recycle system is

$$\tau_F\tau_R s^2 + (\tau_F + \tau_R)s + (1 - K_FK_R) = 0 \tag{2.4}$$

$$\frac{\tau_F\tau_R}{(1 - K_FK_R)} s^2 + \frac{(\tau_F + \tau_R)}{(1 - K_FK_R)} s + 1 = 0 \tag{2.5}$$

This is the standard form of a second-order system, whose time constant is $\sqrt{\tau_F\tau_R/(1 - K_FK_R)}$. As the loop gain in the system K_FK_R (the product of the gains in all units in the forward and recycle path) approaches unity, the time constant of the overall process becomes large. Hence the time constant of an entire process with recycle can be much larger than any of the time constants of its individual units. Figure 2.4 illustrates this for several values of K_FK_R. The value of K_F is constant at unity for these plots, as are the values of τ_F and τ_R. We can see that the effective time constant of the overall process is 25 minutes when $K_R = 0.9$, while the time constants of the individual units are equal to 1 minute. The steady-state gain of the process is $K_F/(1 - K_FK_R)$, so the steady-state effect of the recycle stream also becomes larger as the loop gain approaches unity.

What are the implications of this phenomenon for the plantwide control problem? It means that any change in a recycle process can take a long time to line out back to steady state. We are then tempted not to automate the control loops that handle inventories in recycle loops but rather let the operators manage them. Because the recycle effects are so slow, it is hard to recognize that there is a growing problem in the system inventory. It also takes an equally long time to rectify the situation. Intermediate vessel inventories may overfill or go empty. An imbalance may develop in the inventories of intermediate components. Whenever we do not account for this in the control strategy, the plant's separation section may be subjected to ramplike load disturbances. If the final product column sees this type of disturbance, the product quality controller has difficulty maintaining setpoint. To handle ramp disturbances, special low-frequency-compensated controllers can be used. But these types of controllers are not typically implemented either in conventional control or MPC systems (Belanger and Luyben, 1997). Morud and Skogestad (1996) present a more detailed analysis of the effect of material recycle and heat integration on the dynamic behavior of integrated plants.

2.4.2 Snowball effects

Another interesting observation that has been made about recycle systems is their tendency to exhibit large variations in the magnitude of the recycle flows. Plant operators report extended periods of operation when very small recycle flows occur. It is often difficult to turn the equipment down to such low flowrates. Then, during other periods when feed conditions are not very different, recycle flowrates increase drastically, usually over a considerable period of time. Often the equipment cannot handle such a large load.

We call this high sensitivity of the recycle flowrates to small disturbances the *snowball effect*. We illustrate its occurrence in the simple example below. It is important to note that this is *not* a dynamic effect; it is a *steady-state* phenomenon. But it does have dynamic implications for disturbance propagation and for inventory control. It has nothing to do with closed-loop stability. However, this does not imply that it is independent of the plant's control structure. On the contrary, the extent of the snowball effect is very strongly dependent upon the control structure used.

The large swings in recycle flowrates are undesirable in a plant because they can overload the capacity of the separation section or move the separation section into a flow region below its minimum turndown. Therefore it is important to select a plantwide control structure that avoids this effect. As the example below illustrates and as

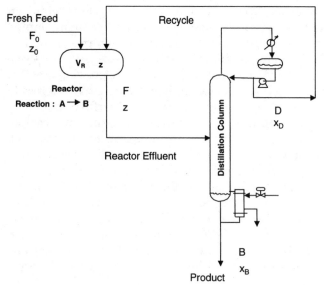

Figure 2.5 Flowsheet of binary recycle process.

more complex processes discussed in later chapters also show, a very effective way to prevent the snowball effect is to apply the following plantwide control heuristic:

A stream somewhere in each liquid recycle loop should be flow controlled.

Let us consider one of the simplest recycle processes imaginable: a continuous stirred tank reactor (CSTR) and a distillation column. As shown in Figure 2.5, a fresh reactant stream is fed into the reactor. Inside the reactor, a first-order isothermal irreversible reaction of component A to produce component B occurs $A \rightarrow B$. The specific reaction rate is k (h^{-1}) and the reactor holdup is V_R (moles). The fresh feed flowrate is F_0 (moles/h) and its composition is z_0 (mole fraction component A). The system is binary with only two components: reactant A and product B. The composition in the reactor is z (mole fraction A). Reactor effluent, with flowrate F (moles/h) is fed into a distillation column that separates unreacted A from product B.

The relative volatilities are such that A is more volatile than B, so the bottoms from the column is the product stream. Its flowrate is B (moles/h) and its composition is x_B (mole fraction A). The amount of A impurity in this product stream is an important control objective and must be maintained at some specified level to satisfy the product quality requirements of the customer.

The overhead distillate stream from the column contains almost all of component A that leaves the reactor because of the purity specifica-

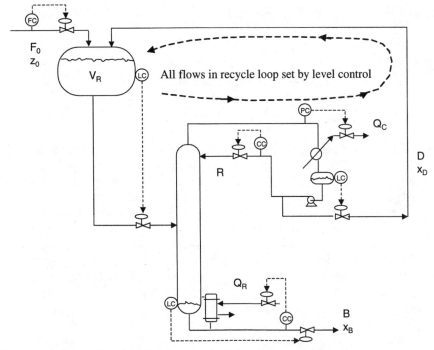

Figure 2.6 Conventional control structure with fixed reactor holdup.

tion on the bottoms stream. It is recycled back to the reactor at a flowrate D and with a composition x_D (mole fraction A). The column has N_T trays and the feed tray is N_F (counting from the bottom). The reflux flowrate is R and the vapor boilup is V (moles/h).

We now explore two alternative control structures for this process.

Conventional control structure. As shown in Fig. 2.6, the following control loops are chosen:

1. Fresh feed flow is controlled.

2. Reactor level is controlled by manipulating reactor effluent flow.

3. Bottoms product purity is controlled by manipulating heat input to the reboiler.

4. Distillate purity is controlled by manipulating reflux flow. Note that we have chosen to use dual composition control (controlling both distillate and bottoms purities) in the distillation column, but there is no *a priori* reason for holding the composition of the recycle stream constant since it does not leave the process. It may be useful to control the composition of this recycle stream for reactor yield pur-

poses or for improved dynamic response. We are often free to find the "best" recycle purity levels in both the design and operation of the plant.

5. Reflux drum level is held by distillate flow (recycle).

6. Base level is held by bottoms flow.

7. Column pressure is controlled by manipulating coolant flowrate to the condenser.

This control scheme is probably what most engineers would devise if given the problem of designing a control structure for this simple plant. Our tendency is to start with setting the flow of the fresh reactant feed stream as the means to regulate plant production rate. We would then work downstream from there as if looking at a steady-state flowsheet and simply connect the recycle stream back to the reactor based upon a standard control strategy for the column.

However, we see in this strategy that there is no flow controller anywhere in the recycle loop. The flows around the loop are set based upon level control in the reactor and reflux drum. Given what we said above, we expect to find that this control structure exhibits the snowball effect. By writing the various overall steady-state mass and component balances around the whole process and around the reactor and column, we can calculate the flow of the recycle stream, at steady state, for any given fresh reactant feed flow and composition. The parameter values used in this specific numerical case are in Table 2.1.

With the control structure in Fig. 2.6 and the base-case fresh feed flow and composition, the recycle flowrate is normally 260.5 moles/h. However, the recycle flow must decrease to 205 moles/h when the fresh feed composition is 0.80 mole fraction A. It must increase to 330 moles/h when the fresh feed compositon changes to pure A. Thus a 25 percent change in the disturbance (fresh feed composition) results in a 60 percent change in recycle flow. With this same control structure and the base-case fresh reactant feed composition, the recycle flow drops to 187 moles/h if the fresh feed flow changes to 215 moles/h. It

TABLE 2.1 Process Data

Base-case fresh feed composition	0.9	mole fraction A
Base-case fresh feed flowrate	239.5	moles/h
Reactor holdup	1250	moles
Reactor effluent flowrate	500	moles/h
Recycle flowrate	260.5	moles/h
Specific reaction rate	0.34086	h^{-1}
Bottoms composition	0.0105	mole fraction A
Recycle composition	0.95	mole fraction A

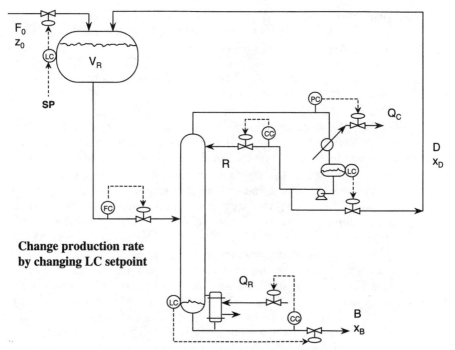

Figure 2.7 Control structure with variable reactor holdup.

must increase to 362 moles/h when the fresh feed flowrate is changed
to 265 moles/h. Thus a 23 percent change in fresh feed flowrate results
in a 94 percent change in recycle flowrate. These snowball effects are
typical for many recycle systems when control structures such as that
shown in Figure 2.6 are used and there is no flow controller somewhere
in the recycle loop.

Variable reactor holdup structure.   An alternative control structure is
shown in Figure 2.7. This strategy differs from the previous one in two
simple but important ways.

1. Reactor effluent flow is controlled.
2. Reactor holdup is controlled by manipulating the fresh reactant
   feed flowrate.

All other control loops are the same. We see here that we cannot change
production rate directly by manipulating the fresh feed flow, because
it is used to control reactor level. However, we must have some means
to set plant throughput, which can be achieved indirectly in this scheme
by changing the setpoint of the reactor level controller. Using the same

numerical case considered previously, the recycle flowrate does not change at all when the fresh feed composition changes. To alter production rate from 215 moles/h to 265 moles/h (a 23 percent change), the reactor holdup must be changed from 1030 moles/h to 1520 moles/h (a 48 percent change). Recycle flow also changes, but only from 285 to 235 moles/h. This is an 18 percent change in recycle flow compared with 94 percent in the alternative strategy.

What are the implications of this phenomenon for the plantwide control problem, when a small disturbance produces a proportionally larger change in recycle flow within the process? Although it is caused by steady-state issues, the snowball effect typically manifests itself in wide dynamic swings in stream flowrates that propagate around the recycle loop. This shows the strong connection between the reaction and separation sections. Whenever all flows in a recycle loop are set by level controllers, wide dynamic excursions occur in these flows because the total system inventory is not regulated. The control system is attempting to control the inventory in each individual vessel by changing the flowrate to its downstream neighbor. In a recycle loop, all level controllers see load disturbances coming from the upstream unit. This causes the flowrate disturbances to propagate around the recycle loop. Thus any disturbance that tends to increase the total inventory in the process (such as an increase in the fresh feed flowrate) will produce large increases in all flowrates around the recycle loop.

2.5   Reaction/Separation Section Interaction

For the process considered in the previous section where the reaction is $A \rightarrow B$, the overall reaction rate depends upon reactor holdup, temperature (rate constant), and reactant composition (mole fraction A) $\mathcal{R} = V_R k z$. The two control structures considered above produce fundamentally different behavior in handling disturbances. In the first, the separation section must absorb almost all of the changes. For example, to increase production rate of component B by 20 percent, the overall reaction rate must increase by 20 percent. Since both reactor temperature (and therefore k) and reactor holdup V_R are held constant, reactor composition z must increase 20 percent. This translates into a very significant change in the composition of the feed stream to the separation section. This means the load on the separation section changes significantly, producing large variations in recycle flowrates.

In the second structure, *both* reactor holdup V_R and reactor composition z can change, so the separation section sees a smaller load disturbance. This reduces the magnitude of the resulting change in recycle flow because the effects of the disturbance can be distributed between the reaction and separation sections.

If the tuning of the reactor level controller in the conventional structure (Fig. 2.6) is modified from normal PI to P only, then changes in production rate also produce changes in reactor holdup. This tends to compensate somewhat for the required changes in overall reaction rate and lessens the impact on the separation section. So both control system structure and the algorithm used in the inventory controller of the reactor affect the amount of this snowball phenomenon.

This example has a liquid-phase reactor, where volume can potentially be varied. If the reactor were vapor phase, reactor volume would be fixed. However, we now have an additional degree of freedom and can vary reactor pressure to affect reaction rate.

We can draw a very useful general conclusion from this simple binary system that is applicable to more complex processes: changes in production rate can be achieved only by changing conditions in the reactor. This means something that affects reaction rate in the reactor must vary: holdup in liquid-phase reactors, pressure in gas-phase reactors, temperature, concentrations of reactants (and products in reversible reactions), and catalyst activity or initiator addition rate. Some of these variables affect the conditions in the reactor more than others. Variables with a large effect are called *dominant*. By controlling the dominant variables in a process, we achieve what is called *partial control*. The term partial control arises because we typically have fewer available manipulators than variables we would like to control. The setpoints of the partial control loops are then manipulated to hold the important economic objectives in the desired ranges.

The plantwide control implication of this idea is that production rate changes should preferentially be achieved by modifying the setpoint of a partial control loop in the reaction section. This means that the separation section will not be significantly disturbed. Using the control structure in Fig. 2.6, changes in production rate require large changes in reactor composition, which disturb the column. Using the control structure shown in Fig. 2.7, changes in production rate are achieved by altering the setpoint of a controlled dominant variable, reactor holdup, with only small changes in reactor composition. This means that the column is not disturbed as much as with the alternative control scheme.

Hence a goal of the plantwide control strategy is to handle variability in production rate and in fresh reactant feed compositions while minimizing changes in the feed stream to the separation section. This may not be physically possible or economically feasible. But if it is, the separation section will perform better to accommodate these changes and to maintain product quality, which is one of the vital objectives for plant operation. Reactor temperature, pressure, catalyst/initiator activity, and holdup are preferred dominant variables to control compared to direct or indirect manipulation of the recycle flows, which of course affect the separation section.

In Chaps. 4 and 6 we discuss specific control issues for chemical reactors and distillation columns. We shall then have much more to say about the important concepts of dominant variables and partial control. Much of the material in those chapters centers on the control of the units individually. However, we also try to show how plantwide control considerations may sometimes alter the control strategy for the unit from what we would normally have in an isolated system.

Some of our previous discussion provides selected clues about why the "best" control structure for an isolated reactor or column may not be the best control strategy when plantwide dynamics are considered.

Let's look again at the simple reactor/column process in Fig. 2.5. In Sec. 2.4.2 we proposed two control structures where both the bottoms composition x_B (the plant product) and the distillate composition x_D (the recycle stream) are controlled, i.e., dual composition control. Bottoms composition must be controlled because it is the product stream leaving the plant and sold to our customers. However, there is *a priori* no reason to control the composition of the recycle stream since this is an internal flow within the plant.

From the perspective of an isolated column, we can achieve better performance in bottoms product composition control by using simple single-end control. Dual composition control means two interacting control loops that normally must be detuned to achieve closed-loop stability. Single-end composition control means one SISO (single-input–single-output) loop that can be tuned up as tightly as the performance/robustness trade-off permits. If we look at just the operation of this distillation column with the control objective to do the best job we can to achieve on-aim product quality, then we would select a single-end control structure for the column.

However, our column is connected via material flow with a reactor. In Chap. 4 we show that reactor control often boils down to two issues: (1) managing energy (temperature control) and (2) keeping as constant as possible the composition and flowrate of the total reactor feed stream (fresh feed plus recycle streams). The latter goal implies that it may in fact be desirable to control the composition of the recycle stream. This minimizes the variablity in recycle impurity composition back into the reactor. This recycle composition is dictated by the economic trade-offs between yield, conversion, energy consumption in the separation section, and reactor size.

Our plantwide control perspective may push us to use a dual composition control system on the column. We would have to loosen up the bottoms composition loop tuning. But smoother reactor operation may reduce disturbances to the column and result in better product quality control.

These are the issues discussed in Part 2: the control of unit operations individually and as part of a plantwide flowsheet.

2.6 Binary System Example

Our simple process considered previously was arbitrarily specified to contain a flowsheet with a reactor, column, and recycle stream. If we step farther back and consider the design of this process, we have many alternative ways to accomplish our objective, which is to take a fresh feed stream containing mostly reactant A and convert it into a stream of mostly product B. In addition to the reactor/column/recycle configuration, we could accomplish the same task by using one large CSTR or by using several CSTRs in series. In this section we analyze these alternatives quantitatively by comparing their steady-state economics (that is, which flowsheet gives the minimum total annual cost considering capital plus energy cost). Then we discuss the dynamic controllability of these alternative flowsheets.

2.6.1 Steady-state design

We neglect the energy cost of cooling the reactor because this will be essentially the same for all alternative flowsheets. Therefore designs with only reactors have to consider just the capital cost of the reactor. Designs with a reactor and column have both energy costs (heat input to the reboiler) and capital costs (reactor, column, reboiler, condenser, and trays). We use here the installed capital costs correlations given by Douglas (1988). The cost of the reactor is assumed to be 5 times the cost of a plain tank. We use a payback period of 3 years to calculate the annual cost of capital.

$$\text{Annual capital cost} = \frac{\text{total capital cost}}{3} \qquad (2.6)$$

Table 2.2 gives equipment sizes and cost data for several alternative designs. Molecular weights are assumed for simplicity to be 50 lb/mole and density is 50 lb/ft^3. An aspect ratio (diameter/length) of 0.5 is used.

TABLE 2.2 Economic Data for CSTRs

Number of CSTRs	1	2	3	4	5
Holdup per vessel, ft^3	59,523	5,802	2,395	1,435	1,009
Diameter, ft	33.6	15.5	11.5	9.7	8.63
Capital cost 10^6 \$	11.8	5.56	4.81	4.66	4.68
Annual capital cost, 10^6 \$/yr	3.95	1.86	1.60	1.55	1.56

TABLE 2.3 Economic Data for CSTR and Stripper

Reactor size, ft^3	800	1000	2500	5000
Reactor diameter, ft	7.98	8.6	11.7	14.7
Trays in stripper	14	14	17	16
Recycle composition, mole fraction A	0.873	0.761	0.391	0.215
Column diameter, ft	8.24	6.04	3.71	3.21
Reboiler energy, 10^6 Btu/h	25.1	13.5	5.09	3.80
Area condenser, ft^2	8360	4496	1697	1267
Area reboiler, ft^2	5020	2698	1018	760
Capital cost, $1000:				
Reactor	810	1090	1645	2535
Column	304	218	152	124
Reboiler	396	264	140	116
Condenser	552	369	196	162
Trays	13	8	5	4
Total capital cost (10^6 $)	2.075	1.949	2.138	2.941
Annual costs, 10^6 $/yr:				
Energy	1.099	0.591	0.223	0.167
Capital	0.692	0.650	0.713	0.980
Total annual cost, 10^6 $/yr	1.79	1.24	0.936	1.15

Additional details of the economic and sizing calculations can be found in Luyben (1993). Notice that the flowsheet with the smallest annual cost has four CSTRs. Now let's compare this system with a process that has one CSTR and a column whose overhead product is recycled back to the reactor. Economic studies of this system have shown that a simple stripping column is cheaper than a full column. Table 2.3 gives size and cost data over a range of reactor sizes.

This simplistic economic evaluation shows that the reactor/stripper process is more economical than the reactors-in-series process. A 2500 ft^3 reactor followed by a stripping column can achieve the same result that would require four 1435 ft^3 reactors in series with no recycle.

In the simple binary process considered above, the 2500 ft^3 reactor with a 17-tray stripper gives the process with the smallest total annual cost: $936,000/yr versus $1,550,000/yr for the best of the CSTR-in-series flowsheets. Thus this process with recycle is more economical, from the viewpoint of steady state, than the alternative process consisting of reactors in series. This is the point we made in Sec. 2.2 about the economic advantage for recycle.

2.6.2 Dynamic controllability

Dynamic simulations of two alternative processes provide a quantitative comparison of their dynamic controllabilities. To strike a balance between simplicity and the economic optimum, we selected the three-CSTR process to compare with the reactor/stripper process. The scheme

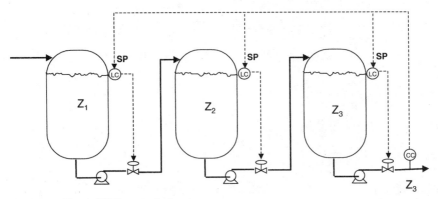

Figure 2.8 Three-CSTR control structure.

used for the three-CSTR process controls the composition of the final product leaving the third reactor (z_3) by changing the setpoint signal to three level controllers for the three vessels (Fig. 2.8). The composition controller has PI action with $K_c = 1$ and $\tau_I = 10.2$ min. A composition transmitter deadtime of 3 minutes is used. Fresh feed is flow-controlled. Level controllers are proportional-only with gains of 10.

The scheme for the reactor/stripper process uses a PI controller to hold product composition (x_B) by manipulating vapor boilup in the stripper. The same analyzer deadtime is used. Proportional level controllers are used for the stripper base (manipulating bottoms flow), the overhead receiver (manipulating recycle flow), and the reactor (manipulating reactor effluent flow) with gains of 2.

Figures 2.9 and 2.10 show the dynamic responses of the two alternative processes for step changes in the fresh feed composition z_0 and fresh feed flowrate F_0. Note the differences in the time scales. The three-CSTR process takes much longer to settle out after the disturbance occurs. However, the maximum deviation of product purity is about half that experienced with the reactor/stripper process. The large holdups in the three reactors filter the disturbances but also slow the process response.

Because the reactor/stripper process is much more attractive economically, it may be the flowsheet of choice despite its larger short-term variability in product quality. This illustrates how plants with recycle are generally more difficult to control than units in series.

2.7 Ternary System Example

We now move on to study another simple process, but again we gain a considerable amount of insight into some important generic concepts for both process design and control (Tyreus and Luyben, 1993). Here

Changes in z_0 (0.9 to 0.7) and F_0 (-20%)

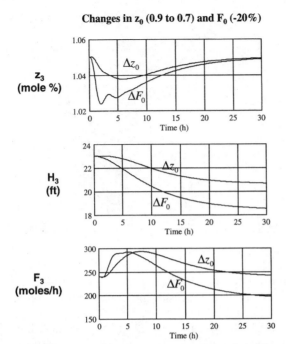

**z_3
(mole %)**

**H_3
(ft)**

**F_3
(moles/h)**

Figure 2.9 Dynamic responses of three-CSTR process.

Changes in z_0 (0.9 to 0.7) and F_0 (-20%)

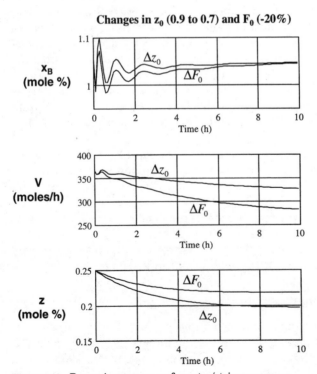

**x_B
(mole %)**

**V
(moles/h)**

**z
(mole %)**

Figure 2.10 Dynamic responses of reactor/stripper process.

we consider a reactor where two reactants A and B form product C: $A + B \rightarrow C$. Since there are three components, we call this system a *ternary example*.

Two kinetic cases will be considered. In the first we assume the reaction rate is so large that the limiting reactant B is completely consumed in the reactor, i.e., there is 100 percent per pass conversion of B. The reactor effluent contains only excess reactant A and product C, so the separation section deals with these two components and recycles A back to the reaction section. In industrial processes, this type of system is typically encountered with extremely hazardous reactants, which we want to be completely consumed in the reactor.

In the second case, which is more general for industrial processes, the reaction rate is not large, so complete one-pass conversion of one reactant would require an excessively large reactor. Economics dictate that reactant concentrations must be significant and recycling of reactants is required. Now the separation section must recover both reactants for recycle.

2.7.1 Complete one-pass reactant conversion

Figure 2.11 shows the ternary process where no B is in the reactor effluent. The size of the reactor and concentrations in the reactor are arbitrary because the consumption of B is independent of these variables. We assume that the separation section consists of a single distillation column. If A is more volatile than C, the overhead product from the column is recycled back to the reactor. If the volatilities are reversed, the bottoms from the column is the recycle stream. Figure 2.11 illustrates the first case.

Two control structures are shown in Fig. 2.11a and b. In both, the composition of component A in the product stream $x_{B,A}$ is controlled by manipulating vapor boilup in the column. This prevents component A from leaving the system. Except for this small amount of A impurity in the product, all A that enters the system must be consumed in the reactor. This illustrates the point we made in Sec. 2.2 about the need to change conditions in the reactor so that the additional reactant is consumed and will not accumulate.

In the first control structure (Fig. 2.11a), both fresh reactant feeds are flow-controlled into the system, with one of the reactants ratioed to the other. This type of control structure is seen quite frequently because we want to set production rate with a reactant feed flow and we know that a stoichiometric ratio of reactants is needed. Unfortunately this strategy does *not* work! It is not possible to feed exactly the stoichiometric amounts of the two reactants. Inaccuracies in flow measurement prevent this from occurring in practice with real instru-

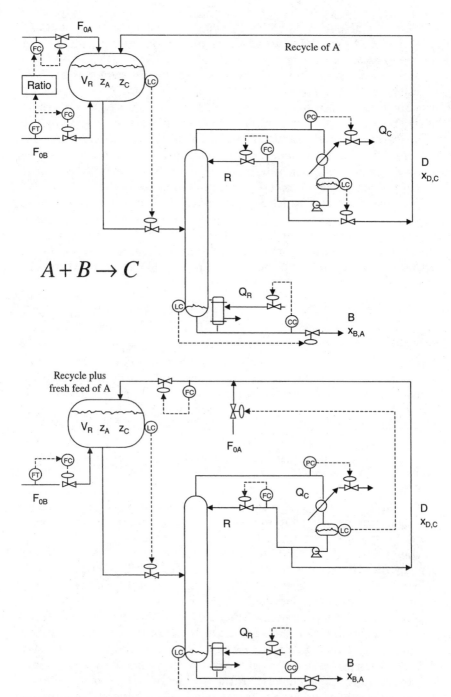

Figure 2.11 Ternary process with complete one-pass conversion of reactant B. (a) Ratio control structure with fixed reactant feed (unworkable); (b) reactant makeup control based on component inventory (workable).

mentation. But even if the flow measurements were perfect, the slightest change in fresh feed compositions would cause the same component imbalance problem. Unless the amounts of the two reactants are perfectly balanced, a gradual buildup will occur of whichever component is in excess. This phenomenon may take hours, days, or weeks. The time depends upon the amount of mismatch between A and B feeding the system.

In the second control structure (Fig. 2.11b), which *does* work, the fresh feed makeup of the limiting reactant (F_{0B}) is flow-controlled. The other fresh feed makeup stream (F_{0A}) is brought into the system to control the liquid level in the reflux drum of the distillation column. The inventory in this drum reflects the amount of A inside the system. If more A is being consumed by reaction than is being fed into the process, the level in the reflux drum will go down. Thus this control structure employs knowledge about the amount of component A in the system to regulate this fresh reactant feed makeup to balance exactly the amount of B fed into the process.

Notice that the total rate of recycle plus fresh feed of A is flow-controlled. There is a flow controller in the recycle loop, which prevents the snowball effect. Sometimes the fresh feed of A is added directly into the reflux drum, making the effect of its flow on reflux drum level more obvious. The piping system where it is not added directly to the drum still gives an immediate effect of makeup flow on drum level because the flowrate of the total stream (recycle plus fresh feed) is held constant. If the fresh feed flow increases, flow from the drum decreases, and this immediately begins to raise the drum level.

2.7.2 Incomplete conversion of both reactants

Now let us consider what is the more common situation where both reactants are present in the reactor effluent. The reaction rate in the reactor $\mathcal{R}$ depends upon the holdup in the reactor V_R, the temperature (through the specific reaction rate k), and the concentrations of both reactants (z_A and z_B):

$$\mathcal{R} = k V_R z_A z_B \tag{2.7}$$

An infinite number of operating conditions in the reactor give exactly the same reaction rate but have different reactor compositions. The only requirement is that the product of the two concentrations (z_A times z_B) be constant. For a given reactor size and temperature, we can have any number of different reactor compositions, and these reactor compositions have a strong impact on the separation system. If z_A is large and z_B is small, there must be a large recycle of A and a small recycle

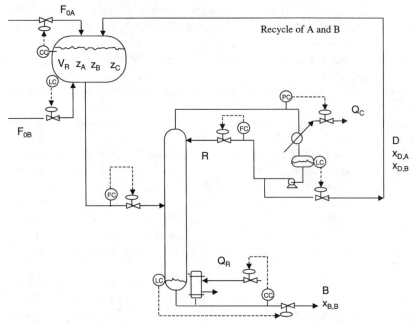

Figure 2.12 Ternary process flowsheet with incomplete conversion of both reactants and one recycle stream.

of B. If the compositions are reversed, the recycle flows are reversed in magnitude. We examine these alternatives later to see how they affect both steady-state economic design and dynamic controllability.

The separation section required to achieve reactant recycle depends upon the relative volatilities of the three components. We consider two cases: (1) the volatility of the product C is heavier or lighter than both of the reactants and (2) the volatility of the product C is intermediate between the reactants. In the first case, we need only one distillation column. In the second, we require two columns if we are limited to a simple two-product configuration.

Single-column case. Let us assume that the relative volatilities are $\alpha_A > \alpha_B > \alpha_C$, so the flowsheet shown in Figure 2.12 is appropriate. Product C is removed from the bottom of the column and contains a small amount of B impurity. It typically has no A because this is the most volatile component. Thus all the A and essentially all the B fed into the process must be consumed in the reactor. The recycle stream is a mixture of mostly B with a modest amount of A and some C. Economics dictate whether this recycle stream should be fairly pure (reducing reactor size but increasing separation costs) or impure.

The control structure shown in Figure 2.12 controls reactor effluent flow to prevent the snowball effect, controls reactor composition by manipulating fresh feed F_{0A}, and controls reactor level with the fresh feed F_{0B}. Both controlled variables are dominant so we have effective partial control of the reactor. This control strategy works. It satisfies the stoichiometry by adjusting the fresh reactant feed flows.

We might be tempted to control reflux drum level with one of the fresh reactant feeds, as done above. The problem with this is that the material in the drum can contain a little of component C mixed with either A or B. Simply looking at the level doesn't tell us anything about component inventories within the process and which might be in excess. The system can fill up with either. Some measure of the composition of at least one of the reactants is required to make this system work. Compositions in the reactor or the recycle stream indicate an imbalance in the amounts of reactants being fed and being consumed. If direct composition measurement is not possible, inferential methods using multiple trays temperatures in the column are sometimes feasible (Yu and Luyben, 1984).

Two-column case. If the relative volatility of the product C is intermediate between the two reactants, a two-column distillation system is typically used. Either the light-out-first (LOF), direct separation sequence, or the heavy-out-first (HOF), indirect separation sequence, can be used. The former is more common because the lightest component only has to be taken overhead once (in the first column) and not twice (as would be the case in the HOF configuration). However there are processes in which the HOF is preferred because it sometimes has the advantage of reducing the exposure of temperature-sensitive components to high base temperatures.

Assuming we use cooling water in both column condensers, the pressure in the first column of the LOF system (with mostly A) will be higher than the pressure in the second column (with mostly B). The base of the first column contains a mixture of B and C, and the base temperature can sometimes be too high for thermally sensitive components. Using the HOF system gives a lower pressure in the first column, and even though the base is now mostly B, the base temperature is sometimes lower than in the LOF system. In addition, component B is being held at high temperature in the base of both columns in the LOF system, and this may be undesirable if B is thermally sensitive.

Whatever separation sequence is chosen, the control structures that work well are quite similar. We will choose the HOF system to illustrate this type of process. Figure 2.13 gives a sketch of a ternary process with two recycle streams. The heaviest component B is recycled back to the reactor from the base of the first column. The lightest component A is recycled back to the reactor from the top of the second column.

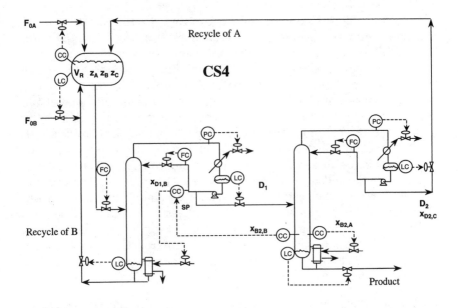

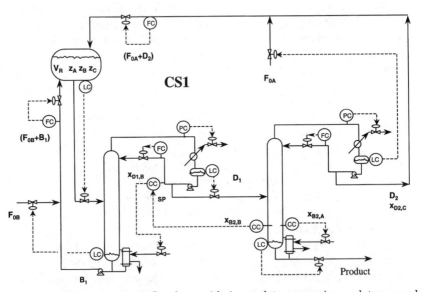

Figure 2.13 Ternary process flowsheet with incomplete conversion and two recycle streams (heavy-out-first sequence). (*a*) Control structure CS4: reactor composition and level control (workable); (*b*) control structure CS1: reactant makeup control based on component inventories (workable).

Figure 2.13a and b shows two control structures that work (CS4 and CS1). Both of these provide a mechanism for adjusting the fresh feed reactant flowrates so that the overall stoichiometry can be satisfied. In CS4 this is accomplished by measuring reactor composition. In CS1 it is accomplished by deducing the amounts of the reactants in the process from two levels in the two recycle loops.

In both strategies the control of the separation section is similar:

1. In both columns, reflux flows are fixed (or ratioed to feedrates) and pressures are controlled by condenser cooling.

2. The impurity of A $(x_{B2,A})$ in the product stream B_2 from the second column is controlled by vapor boilup in the second column.

3. The impurity of B $(x_{B2,B})$ in the product stream B_2 from the second column is controlled by vapor boilup in the first column through a composition-composition cascade control system. Any B that goes overhead in the first column comes out the bottom of the second column. So the first column must be operated to prevent B from going overhead. The impurity of B in the first column distillate $(x_{D1,B})$ is controlled by a composition controller that manipulates the vapor boilup in the first column. The setpoint of this composition controller is changed by a second composition controller looking at the impurity of B in the product stream $(x_{B2,B})$.

Control structure CS4 (Fig. 2.13a) controls reactor effluent flow, brings fresh A in to hold reactor composition z_A, and brings fresh B in to control reactor level. In both columns, the base levels are controlled by manipulating bottoms flowrates and the reflux drum levels are controlled by manipulating distillate flowrates.

Control structure CS1 (Fig. 2.13b) controls the flowrates of the two total light and heavy recycle streams; i.e., the sum of the fresh feed and recycle of A $(F_{0A} + D_2)$ is flow-controlled and the sum of the fresh feed and recycle of B $(F_{0B} + B_1)$ is flow-controlled. The fresh reactant A feed controls the level in the reflux drum of the second column, which reflects this component's inventory within the process. Similarly, the fresh reactant B feed controls the level in the base of the first column.

Both of these control structures have the slight disadvantage of lacking a single direct handle to set production rate, i.e., a one-to-one relationship with product flow. Desired throughput must be achieved by changing the setpoint of the reactor concentration controller, the reactor level controller, the reactor effluent flow controller, and/or the recycle flow controllers (one or both). Structure CS4 has another disadvantage since it requires a composition measurement, which can be very expensive and unreliable in many systems.

We could easily propose many other control structures for this pro-

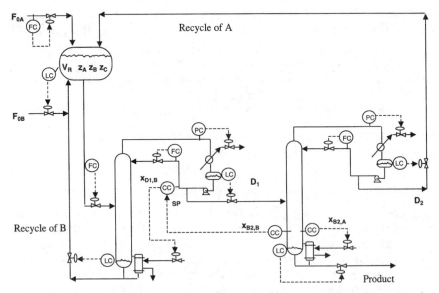

Figure 2.14 Ternary process flowsheet with incomplete conversion and two recycle streams (heavy-out-first sequence): control structure CS2 with fixed flow of one reactant (unworkable).

cess, but most *do not work* in these types of systems. Schemes where one of the reactant fresh feeds is simply flow-controlled into the process do not work unless the per-pass conversion of this limiting component is quite high; i.e., the concentration of this component in the reactor effluent is very small. An analysis of this problem is given in Luyben et al. (1996).

For example, consider the control system shown in Figure 2.14. Here there is a direct handle on production: the flow of fresh A into the system. However, this scheme does not work. Figure 2.15a illustrates that the system is able to handle a very small (2 percent) change in fresh feed flow. But if the change in fresh feed flow is increased to 5 percent, the system fills up with A and shuts down after 150 hours (see Fig. 2.15b). If the increase is +10 percent (Fig. 2.15c), the system shuts down in 70 hours. Thus this control structure can handle only very small disturbances. The imbalance in chemical components and the long time period over which the problem occurs highlight the importance of these phenomena in the plantwide control problem.

2.7.3 Stability analysis

To gain some understanding of what is happening in the results shown in Fig. 2.15 and to explain why the process shuts down, it is useful to

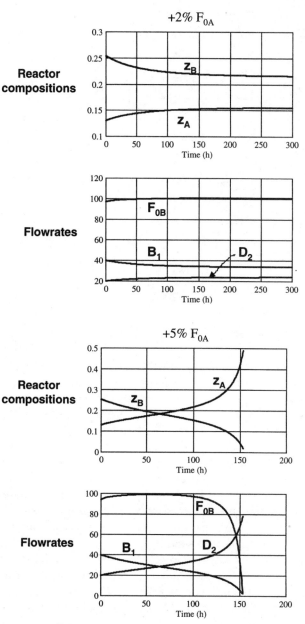

Figure 2.15 Dynamic response of ternary process with CS2 for change in fixed reactant feed rate. (a) 2 percent increase; (b) 5 percent increase; (c) 10 percent increase.

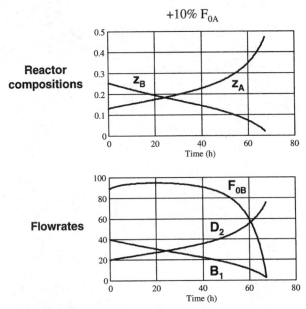

Figure 2.15   *(Continued)*

look at a simple model of the process and to see what such a model predicts concerning the stability of the process. The results in Fig. 2.15 show that the disturbance in F_{0A} drives the reactor compositions into a region where z_A becomes larger than z_B and then a shutdown eventually occurs after several hours.

Let us derive a dynamic model of the process with control structure CS2 included. A rigorous model of the reactor and the two distillation columns would be quite complex and of very high order. Because the dynamics of the liquid-phase reactor are much slower than the dynamics of the separation section in this process, we can develop a simple second-order model by assuming the separation section dynamics are instantaneous. Thus the separation section is always at steady state and is achieving its specified performance, i.e., product and recycle purities are at their setpoints. Given a flowrate F and the composition z_A/z_B of the reactor effluent stream, the flowrates of the light and heavy recycle streams D_2 and B_1 can be calculated from the algebraic equations

$$D_2 = \frac{Fz_A - A_{\text{loss}}}{x_{D2,A}} \tag{2.8}$$

$$B_1 = \frac{Fz_B - B_{\text{loss}}}{x_{B1,B}} \tag{2.9}$$

where A_{loss}, B_{loss} = small molar flowrates of components A and B in
product stream B_2 leaving base of second column
(assumed to be constant)

$x_{D2,A}$, $x_{B1,B}$ = purities of recycle streams

Perfect reactor level control is assumed. The reactor effluent flowrate F is fixed in control structure CS2. The two state variables of the system are the two reactor compositions z_A and z_B. The two nonlinear ordinary differential equations describing the system are

$$V_R \frac{dz_A}{dt} = F_{0A} + D_2 x_{D2,A} - F z_A - V_R k z_A z_B \qquad (2.10)$$

$$V_R \frac{dz_B}{dt} = F_{0B} + B_1 x_{B1,B} - F z_B - V_R k z_A z_B \qquad (2.11)$$

At any point in time we know z_A and z_B. The variables F, F_{0A}, k, V_R, A_{loss}, B_{loss}, $x_{D2,A}$, and $x_{B1,B}$ are constant. At each point in time Eqs. (2.8) and (2.9) can be used to find the recycle flowrates. A total molar balance around the reactor can be used to calculate the makeup flowrate of component B, F_{0B}. Remember that the reaction is $A + B \rightarrow C$, so moles are not conserved.

$$F_{0A} + F_{0B} + D_2 + B_1 = F + V_R k z_A z_B \qquad (2.12)$$

The two nonlinear ordinary differential equations can be linearized around the steady-state values of the reactor compositions $\bar{z}_A$ and $\bar{z}_B$. Laplace transforming gives the characteristic equation of the system. It is important to remember that we are looking at the closed-loop system with control structure CS2 in place. Therefore Eq. (2.13) is the closed-loop characteristic equation of the process:

$$s^2 + s \left[k\bar{z}_B + \frac{F}{V_R x_{B1,B}} \right] + \frac{kF}{V_R} \left[\frac{\bar{z}_B}{x_{B1,B}} - \frac{\bar{z}_A}{x_{D2,A}} \right] = 0 \qquad (2.13)$$

Thus the linear analysis predicts that the system will be closed-loop unstable when

$$\frac{\bar{z}_B}{x_{B1,B}} < \frac{\bar{z}_A}{x_{D2,A}} \qquad (2.14)$$

If the two recycle purities are about the same ($x_{B1,B} \cong x_{D2,A}$), which is the case in the numerical example considered earlier in the chapter, the linear analysis predicts that instability will occur when z_A is bigger than z_B. This is exactly what we observed in Fig. 2.15.

The physical reason for this instability is the lack of some mechanism in the process or in the control structure to ensure that the A and B component balances are satisfied in this integrating plantwide process. Both reactant components are prevented from leaving the system by the impurity controllers that are looking at the product stream. Thus essentially all of the reactants fed into the system must be consumed by chemical reaction. And the stoichiometry must be satisfied down to the last molecule: every mole of A requires exactly one mole of B to react with. The flowrates of the fresh feed cannot be controlled in an open-loop fashion anywhere nearly accurately enough to match the molecules of the two reactants exactly. This is why we need some information about the amounts of the two components in the system. This knowledge can be used in a feedback control system to make some adjustments so that the component in excess does not continue to build up in the system.

2.7.4   Modification of CS2

Both of the control structures discussed in Sec. 2.7.2 (CS1 and CS4) work because they detect the inventories of the reactant components A and B in the system and bring in fresh feed streams to balance the consumption of the two components. Structure CS1 does this by using the liquid level in the reflux drum of the second column as an indicator of the amount of A in the system and the liquid level in the base of the first column as an indicator of the amount of B in the system. Structure CS4 uses a composition analyzer to measure directly the concentration of one of the reactants in the reactor. But both of these structures lack a direct handle on production rate.

Control structure CS2 has such a direct handle, but this structure does not work. However, a modification can be made to CS2 that will make it work. The basic idea is to recognize that the separation section acts like an on-line analyzer. Any component B in the reactor effluent gets recycled in B_1. Any component A in the reactor effluent gets recycled in D_2. Therefore, the flowrates of these two streams give a direct indication of the amounts of the two reactants in the system.

Figure 2.16 shows a control scheme in which the ratio of the two recycle flowrates is controlled by adjusting the flowrate of the reactor effluent. The dynamics of the separation system must be considered because a change in the amount of A in the reactor effluent has to work its way through two columns before showing up as a change in the flowrate of D_2. Thus a lag is added to the measurement of B_1 before it is used to calculate the ratio. This control structure works.

In this modified CS2, the feedback adjustment that is made to adjust for any imbalance in the amounts of the two reactants in the system is a change in the reactor effluent flowrate to achieve a constant ratio

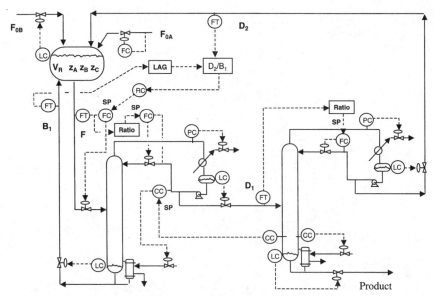

Figure 2.16 Ternary process flowsheet with incomplete conversion and two recycle streams (heavy-out-first sequence): control structure CS2C using separation as analyzer for control of D_2/B_1 ratio (workable).

of the two recycle flowrates. This works because these flowrates give good indications of the concentrations of the two reactants in the reactor. The two columns act like composition analyzers, separating the A and B components from the product C.

In the numerical case studied in this chapter, we considered a liquid-phase reactor with dynamics that were slower than the dynamics of the separation system. Suppose we have a process with a vapor-phase reactor whose dynamics are much faster than those of the separation section. Will the modified CS2 control structure work in this process?

Luyben et al. (1996) explored this question in detail by developing a rigorous simulation of such a process. Their results demonstrate that the proposed control structure does provide effective control for processes with fast reactor dynamics. The time constant of the separation section is about 30 minutes. The reactor time constant was reduced to 3 minutes, and control was still good.

2.7.5   Reactor composition trade-offs

As discussed earlier, if the concentration of A (or B) in the reactor is essentially zero, we can flow-control the fresh feed of A (or B) into the system, and large disturbances can be handled. In the numerical case

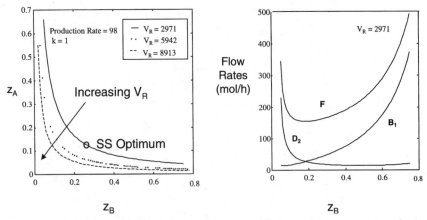

Figure 2.17 Steady-state design for ternary process with incomplete conversion and two recycle streams (heavy-out-first sequence).

presented in the previous section, the steady-state economic design of the process yielded reactor compositions that are $z_A = 0.15$ mole fraction and $z_B = 0.25$ mole fraction. It is cheaper to recycle B than A because B comes out the bottom of the first column and does not have to be vaporized. Component A, on the other hand, must be vaporized twice as it is taken overhead in both columns. Therefore the steady-state separation design favors smaller z_A and larger z_B. But remember that if reactor temperature and holdup are constant (fixed k and V_R), the product of the two concentrations must be fixed to achieve a given production rate of C.

Figure 2.17 illustrates that we must lie somewhere on the hyperbolic line in the $z_A - z_B$ plane. At any position on one of the constant reactor volume lines, the production rate is constant. The concentrations fed to the separation section vary with our choice of location on this curve. For large z_A and small z_B, the recycle of A (D_2) is large. For large z_B and small z_A, the recycle of B (B_1) is large.

Since we are dealing with the product of the two reactant concentrations, making them approximately equal is the best way to minimize reactor holdup. Thus steady-state reactor design favors compositions that are somewhat similar. From a dynamic viewpoint, the system can handle disturbances more easily if the concentrations of the two reactants are very different (very small z_A and large z_B). We saw an indication of this in the ternary process considered earlier. Control structure CS2 worked when the concentration of the limiting reactant was very low, but failed when the concentration of the limiting reactant was in the 0.15 mole fraction region.

So this simple process provides another nice example of the very common situation where a conflict exists between steady-state economic design and dynamic controllability.

2.8 Conclusion

In this chapter we have looked at the steady-state and dynamic implications of integrated processes with recycle (as compared with units connected only in series). From a dynamic standpoint we found that recycles provide positive feedback that alters the overall time constant of the process. From a steady-state viewpoint, recycles introduce the possibility of the snowball effect, where a small change in throughput or feed composition can produce a large change in recycle flowrates. These features restrict the set of workable control structures for an integrated process. Several simple processes were used to illustrate the interaction between the reaction and separation sections. The generic conclusion was to control dominant variables using local manipulators in the reaction section. We then achieve production rate changes by manipulating the setpoints so that disturbances to the separation section are minimized, thereby reducing product quality variability. Another point that was highlighted involved the need for the control strategy to account for the chemical component balances, i.e., to keep track of the inventory of components within the system.

2.9 References

Belanger, P. W., and Luyben, W. L. "Design of Low-Frequency Compensators for Improvement of Plantwide Regulatory Performance," *Ind. Eng. Chem. Res.*, **36**, 5339–5347 (1997).

Douglas, J. M. *Conceptual Design of Chemical Processes*, New York: McGraw-Hill (1988).

Luyben, W. L. "Dynamics and Control of Recycle Systems: 2. Comparison of Alternative Process Designs," *Ind. Eng. Chem. Res.*, **32**, 476–486 (1993).

Luyben, M. L., Tyreus, B. D., Luyben, W. L. "Analysis of Control Structures for Reaction/ Separation/ Recycle Processes with Second-Order Reactions," *Ind. Eng. Chem. Res.*, **35**, 758–771 (1996).

Morud, J., and Skogestad, S. "Dynamic Behavior of Integrated Plants," *J. Proc. Cont.*, **6**, 145–156 (1996).

Terrill, D. L., and Douglas, J. M. "Heat-Exchanger Network Analysis. 1. Optimization," *Ind. Eng. Chem. Res.*, **26**, 685–691 (1987).

Tyreus, B. D., and Luyben, W. L. "Dynamic and Control of Recycle Systems: 4. Ternary Systems with One or Two Recycle Streams," *Ind. Eng. Chem. Res.*, **32**, 1154–1162 (1993).

Yu, C. C., and Luyben, W. L. "Use of Multiple Temperatures for the Control of Multicomponent Distillation Columns," *Ind. Eng. Chem. Proc. Des. Dev.*, **23**, 590–597 (1984).

3

Plantwide Control Design Procedure

3.1 Introduction

In an industrial environment, a plant's control strategy should be simple enough, at least conceptually, so that everyone from the operator to the plant manager can understand how it works. Our governing philosophy is *it is always best to utilize the simplest control system that will achieve the desired objectives*. The more complex the process, the more desirable it is to have a simple control strategy. This view differs radically from much of the current academic thinking about process control, which suggests that a complex process demands complex control. Our viewpoint is a result of many years of working on practical plant control problems, where it is important to be able to identify whether an operating problem has its source in the process or in the control system.

The goals for an effective plantwide process control system include (1) safe and smooth process operation; (2) tight control of product quality in the face of disturbances; (3) avoidance of unsafe process conditions; (4) a control system run in automatic, not manual, requiring minimal operator attention; (5) rapid rate and product quality transitions; and (6) zero unexpected environmental releases.

As illustrated in the previous chapter, the need for a plantwide control perspective arises from three important features of integrated processes: the effects of material recycle, of chemical component inventories, and of energy integration. We have shown several control strategies that highlight important general issues. However, we did not describe how we arrived at these strategies, and many of our choices may seem mysterious at this point. Why, for instance, did we choose to use fresh liquid reactant feed streams in the control of liquid inventories? What prompted us to have a reactor composition analyzer? Why were we concerned with a single direct handle to set production rate?

In this chapter we outline the nine basic steps of a general heuristic plantwide control design procedure (Luyben et al., 1997). After some preliminary discussion of the fundamentals on which this procedure is based, we outline each step in general terms. We also summarize our justification for the sequence of steps. The method is illustrated in applications to four industrial process examples in Part 3.

The procedure essentially decomposes the plantwide control problem into various levels. It forces us to focus on the unique features and issues associated with a control strategy for an entire plant. We highlighted some of these questions in Chap. 1 in discussing the HDA process. How do we manage energy? How is production rate controlled? How do we control product quality? How do we determine the amounts of fresh reactants to add?

Our plantwide control design procedure (Fig. 3.1) satisfies the two fundamental chemical engineering principles, namely the overall conservation of energy and mass. Additionally, the procedure accounts for nonconserved entities within a plant such as chemical components (produced and consumed) and entropy (produced). In fact, five of the nine steps deal with plantwide control issues that would not be addressed by simply combining the control systems from all of the individual unit operations.

Steps 1 and 2 establish the objectives of the control system and the available degrees of freedom. Step 3 ensures that any production of heat (entropy) within the process is properly dissipated and that the propagation of thermal disturbances is prevented. In Steps 4 and 5 we

1. Establish Control Objectives
2. Determine Control Degrees of Freedom
3. Establish Energy Management System
4. Set Production Rate
5. Control Product Quality and Handle Safety, Environmental, and Operational Constraints
6. Fix a Flow in Every Recycle Loop and Control Inventories (Pressures and Liquid Levels)
7. Check Component Balances
8. Control Individual Unit Operations
9. Optimize Economics and Improve Dynamic Controllability

Figure 3.1 Nine steps of plantwide control design procedure.

satisfy the key business objectives concerning production rate, product quality, and safety. Step 6 involves total mass balance control, whereas in Step 7 we ensure that nonconserved chemical components are accounted for. That concludes the plantwide control aspects. In Step 8 we complete the control systems for individual unit operations. Finally, Step 9 uses the remaining degrees of freedom for optimization and improved dynamic controllability. This heuristic procedure will generate a workable plantwide control strategy, which is not necessarily the *best* solution. Because the design problem is open-ended, the procedure will not produce a unique solution.

The plantwide control design procedure presented here was developed after many years of work and research in the fields of process control and process design. Research efforts by a number of people in industry and at universities have contributed essential ideas and concepts. We have assembled, analyzed, and processed this prior work to reach a logical, coherent, step-by-step procedure. We want to acknowledge these previous contributions and state that we are indeed fortunate to stand upon the shoulders of many giants. Listed below are some of the fundamental concepts and techniques that form the basis of the procedure.

3.2 Basic Concepts of Plantwide Control

3.2.1 Buckley basics

Page Buckley (1964), a true pioneer with DuPont in the field of process control, was the first to suggest the idea of separating the plantwide control problem into two parts: material balance control and product quality control. He suggested looking first at the flow of material through the system. A logical arrangement of level and pressure control loops is established, using the flowrates of the liquid and gas process streams. No controller tuning or inventory sizing is done at this step. The idea is to establish the inventory control system by setting up this "hydraulic" control structure as the first step.

He then proposed establishing the product-quality control loops by choosing appropriate manipulated variables. The time constants of the closed-loop product-quality loops are estimated. We try to make these as small as possible so that good, tight control is achieved, but stability constraints impose limitations on the achieveable performance.

Then the inventory loops are revisited. The liquid holdups in surge volumes are calculated so that the time constants of the liquid level loops (using proportional-only controllers) are a factor of 10 larger than the product-quality time constants. This separation in time constants permits independent tuning of the material-balance loops and the product-quality loops. Note that most level controllers should be proportional-only (P) to achieve flow smoothing.

3.2.2 Douglas doctrines

Jim Douglas (1988) of the University of Massachusetts has devised a hierarchical approach to the conceptual design of process flowsheets. Although he primarily considers the steady-state aspects of process design, he has developed several useful concepts that have control structure implications.

Douglas points out that in the typical chemical plant the costs of raw materials and the value of the products are usually much greater than the costs of capital and energy. This leads to the two *Douglas doctrines:*

1. Minimize losses of reactants and products.

2. Maximize flowrates through gas recycle systems.

The first idea implies that we need tight control of stream compositions exiting the process to avoid losses of reactants and products. The second rests on the principle that yield is worth more than energy. Recycles are used to improve yields in many processes, as was discussed in Chap. 2. The economics of improving yields (obtaining more desired products from the same raw materials) usually outweigh the additional energy cost of driving the recycle gas compressor.

The control structure implication is that we do not attempt to regulate the gas recycle flow and we do not worry about what we control with its manipulation. We simply maximize its flow. This removes one control degree of freedom and simplifies the control problem.

3.2.3 Downs drill

Jim Downs (1992) of Eastman Chemical Company has insightfully pointed out the importance of looking at the chemical component balances around the entire plant and checking to see that the control structure handles these component balances effectively. The concepts of overall component balances go back to our first course in chemical engineering, where we learned how to apply mass and energy balances to any system, microscopic or macroscopic. We did these balances for individual unit operations, for sections of a plant, and for entire processes.

But somehow these basics are often forgotten or overlooked in the complex and intricate project required to develop a steady-state design for a large chemical plant and specify its control structure. Often the design job is broken up into pieces. One person will design the reactor and its control system and someone else will design the separation section and its control system. The task sometimes falls through the cracks to ensure that these two sections operate effectively when coupled together. Thus it is important that we perform the *Downs drill*.

We must ensure that all components (reactants, products, and inerts) have a way to leave or be consumed within the process. The consideration of inerts is seldom overlooked. Heavy inerts can leave the system in the bottoms product from a distillation column. Light inerts can be purged from a gas recycle stream or from a partial condenser on a column. Intermediate inerts must also be removed in some way, for example in sidestream purges or separate distillation columns.

Most of the problems occur in the consideration of reactants, particularly when several chemical species are involved. All of the reactants fed into the system must either be consumed via reaction or leave the plant as impurities in the exiting streams. Since we usually want to minimize raw material costs and maintain high-purity products, most of the reactants fed into the process must be chewed up in the reactions. And the stoichiometry must be satisfied *down to the last molecule*.

Chemical plants often act as pure integrators in terms of reactants. This is due to the fact that we prevent reactants from leaving the process through composition controls in the separation section. Any imbalance in the number of moles of reactants involved in the reactions, no matter how slight, will result in the process gradually filling up with the reactant component that is in excess. The ternary system considered in Chap. 2 illustrated this effect. There must be a way to adjust the fresh feed flowrates so that exactly the right amounts of the two reactants are fed in.

3.2.4 Luyben laws

Three laws have been developed as a result of a number of case studies of many types of systems:

1. A stream somewhere in all recycle loops should be flow controlled. This is to prevent the snowball effect and was discussed in Chap. 2.

2. A fresh reactant feed stream cannot be flow-controlled unless there is essentially complete one-pass conversion of one of the reactants. This law applies to systems with reaction types such as $A + B \rightarrow$ products and was discussed in Chap. 2. In systems with consecutive reactions such as $A + B \rightarrow M + C$ and $M + B \rightarrow D + C$, the fresh feeds can be flow-controlled into the system because any imbalance in the ratios of reactants is accommodated by a shift in the amounts of the two products (M and D) that are generated. An excess of A will result in the production of more M and less D. An excess of B results in the production of more D and less M.

3. If the final product from a process comes out the top of a distillation column, the column feed should be liquid. If the final product comes out the bottom of a column, the feed to the column should be vapor (Cantrell et al., 1995). Changes in feed flowrate or feed composition have less of a dynamic effect on distillate composition than they do on bottoms composition if the feed is saturated liquid. The reverse is true if the feed is saturated vapor: bottoms is less affected than distillate. If our primary goal is to achieve tight product quality control, the basic column design should consider the dynamic implications of feed thermal conditions. Even if steady-state economics favor a liquid feed stream, the profitability of an operating plant with a product leaving the bottom of a column may be much better if the feed to the column is vaporized. This is another example of the potential conflict between steady-state economic design and dynamic controllability.

3.2.5 Richardson rule

Bob Richardson of Union Carbide suggested the heuristic that the largest stream should be selected to control the liquid level in a vessel. This makes good sense because it provides more *muscle* to achieve the desired control objective. An analogy is that it is much easier to maneuver a large barge with a tugboat than with a life raft. We often use the expression that you can't make a garbage truck drive like a Ferrari. But this is not necessarily true. If you put a 2000-hp engine in the garbage truck (and redesigned the center of gravity), you could make it handle just like a sports car. The point is that the bigger the handle you have to affect a process, the better you can control it. This is why there are often fundamental conflicts between steady-state design and dynamic controllability.

3.2.6 Shinskey schemes

Greg Shinskey (1988), over the course of a long and productive career at Foxboro, has proposed a number of "advanced control" structures that permit improvements in dynamic performance. These schemes are not only effective, but they are simple to implement in basic control instrumentation. Liberal use should be made of ratio control, cascade control, override control, and valve-position (optimizing) control. These strategies are covered in most basic process control textbooks.

3.2.7 Tyreus tuning

One of the vital steps in developing a plantwide control system, once both the process and the control structure have been specified, is to determine the algorithm to be used for each controller (P, PI, or PID) and to tune each controller. We strongly recommend the use of P-only controllers for liquid levels (even in some liquid reactor applications). Tuning of a P controller is usually trivial: set the controller gain equal to 1.67. This will have the valve wide open when the level is at 80 percent and the valve shut when the level is at 20 percent (assuming the stream flowing out of the vessel is manipulated to control liquid level; if the level is controlled by the inflowing stream the action of the controller is reverse instead of direct).

For other control loops, we suggest the use of PI controllers. The relay-feedback test is a simple and fast way to obtain the ultimate gain (K_u) and ultimate period (P_u). Then either the Ziegler-Nichols settings (for very tight control with a closed-loop damping coefficient of about 0.1) or the Tyreus-Luyben (1992) settings (for more conservative loops where a closed-loop damping coefficient of 0.4 is more appropriate) can be used:

$$K_{ZN} = K_u/2.2 \qquad \tau_{ZN} = P_u/1.2$$

$$K_{TL} = K_u/3.2 \qquad \tau_{TL} = 2.2P_u$$

The use of PID controllers should be restricted to those loops where two criteria are both satisfied: the controlled variable should have a very large signal-to-noise ratio and tight dynamic control is really essential from a feedback control stability perspective. The classical example of the latter is temperature control in an irreversible exothermic chemical reactor (see Chap. 4).

3.3   Steps of Plantwide Process Control Design Procedure

In this section we discuss each step of the design procedure in detail.

Step 1: Establish control objectives

Assess the steady-state design and dynamic control objectives for the process.

This is probably the most important aspect of the problem because different control objectives lead to different control structures. There is an old Persian saying "If you don't know where you are going, any road will get you there!" This is certainly true in plantwide control. The "best" control structure for a plant depends upon the design and control criteria established.

These objectives include reactor and separation yields, product quality specifications, product grades and demand determination, environmental restrictions, and the range of safe operating conditions.

Step 2: Determine control degrees of freedom

Count the number of control valves available.

This is the number of degrees of freedom for control, i.e., the number of variables that can be controlled to setpoint. The valves must be legitimate (flow through a liquid-filled line can be regulated by only one control valve). The placement of these control valves can sometimes be made to improve dynamic performance, but often there is no choice in their location.

Most of these valves will be used to achieve basic regulatory control of the process: (1) set production rate, (2) maintain gas and liquid inventories, (3) control product qualities, and (4) avoid safety and environmental constraints. Any valves that remain after these vital tasks have been accomplished can be utilized to enhance steady-state economic objectives or dynamic controllability (e.g., minimize energy consumption, maximize yield, or reject disturbances).

During the course of the subsequent steps, we may find that we lack suitable manipulators to achieve the desired economic control objectives. Then we must change the process design to obtain additional handles. For example, we may need to add bypass lines around heat exchangers and include auxiliary heat exchangers.

Step 3: Establish energy management system

Make sure that energy disturbances do not propagate throughout the process by transferring the variability to the plant utility system.

We use the term *energy management* to describe two functions: (1) We must provide a control system that removes exothermic heats of reaction from the process. If heat is not removed to utilities directly at the reactor, then it can be used elsewhere in the process by other unit operations. This heat, however, must ultimately be dissipated to utilities. (2) If heat integration does occur between process streams, then the second function of energy management is to provide a control system that prevents the propagation of thermal disturbances and ensures the exothermic reactor heat is dissipated and not recycled. Process-to-process heat exchangers and heat-integrated unit operations must be analyzed to determine that there are sufficient degrees of freedom for control.

Heat removal in exothermic reactors is crucial because of the potential for thermal runaways. In endothermic reactions, failure to add enough heat simply results in the reaction slowing up. If the exothermic reactor is running adiabatically, the control system must prevent excessive temperature rise through the reactor (e.g., by setting the ratio of the flowrate of the limiting fresh reactant to the flowrate of a recycle stream acting as a thermal sink). More details of reactor control are discussed in Chap. 4.

Heat transfer between process streams can create significant interaction. In the case of reactor feed/effluent heat exchangers it can lead to positive feedback and even instability. Where there is partial condensation or partial vaporization in a process-to-process heat exchanger, disturbances can be amplified because of heat of vaporization and temperature effects.

For example, suppose the temperature of a stream being fed to a distillation column is controlled by manipulating steam flowrate to a feed preheater. And suppose the stream leaving the preheater is partially vaporized. Small changes in composition can result in very large changes in the fraction of the stream that is vaporized (for the same pressure and temperature). The resulting variations in the liquid and vapor rates in the distillation column can produce severe upsets.

Heat integration of a distillation column with other columns or with reactors is widely used in chemical plants to reduce energy consumption. While these designs look great in terms of steady-state economics, they can lead to complex dynamic behavior and poor performance due to recycling of disturbances. If not already included in the design, trim heaters/coolers or heat exchanger bypass lines must be added to prevent this. Energy disturbances should be transferred to the plant utility system whenever possible to remove this source of variability from the process units. Chapter 5 deals with heat exchanger systems.

Step 4: Set production rate

Establish the variables that dominate the productivity of the reactor and determine the most appropriate manipulator to control production rate.

Throughput changes can be achieved only by altering, either directly or indirectly, conditions in the reactor. To obtain higher production rates, we must increase overall reaction rates. This can be accomplished by raising temperature (higher specific reaction rate), increasing reactant concentrations, increasing reactor holdup (in liquid-phase reactors), or increasing reactor pressure (in gas-phase reactors).

Our first choice for setting production rate should be to alter one of these variables in the reactor. The variable we select must be dominant for the reactor. Dominant reactor variables always have significant effects on reactor performance. For example, temperature is often a dominant reactor variable. In irreversible reactions, specific rates increase exponentially with temperature. As long as reaction rates are not limited by low reactant concentrations, we can *increase* temperature to increase production rate in the plant. In reversible exothermic reactions, where the equilibrium constant decreases with increasing temperature, reactor temperature may still be a dominant variable. If the reactor is large enough to reach chemical equilibrium at the exit, we can *decrease* reactor temperature to increase production.

There are situations where reactor temperature is not a dominant variable or cannot be changed for safety or yield reasons. In these cases, we must find another dominant variable, such as the concentration of the limiting reactant, flowrate of initiator or catalyst to the reactor, reactor residence time, reactor pressure, or agitation rate.

Once we identify the dominant variables, we must also identify the manipulators (control valves) that are most suitable to control them. The manipulators are used in feedback control loops to hold the dominant variables at setpoint. The setpoints are then adjusted to achieve the desired production rate, in addition to satisfying other economic control objectives.

Whatever variable we choose, we would like it to provide smooth and stable production rate transitions and to reject disturbances. We often want to select a variable that has the least effect on the separation section but also has a rapid and direct effect on reaction rate in the reactor without hitting an operational constraint.

When the setpoint of a dominant variable is used to establish plant production rate, the control strategy must ensure that the right amounts of fresh reactants are brought into the process. This is often accomplished through fresh reactant makeup control based upon liquid levels or gas pressures that reflect component inventories. We must keep these ideas in mind when we reach Steps 6 and 7.

However, design constraints may limit our ability to exercise this strategy concerning fresh reactant makeup. An upstream process may establish the reactant feed flow sent to the plant. A downstream process may require on-demand production, which fixes the product flowrate from the plant. In these cases, the development of the control strategy becomes more complex because we must somehow adjust the setpoint of the dominant variable on the basis of the production rate that has been specified externally. We must balance production rate with what has been specified externally. This cannot be done in an open-loop sense. Feedback of information about actual internal plant conditions is required to determine the accumulation or depletion of the reactant components. This concept was nicely illustrated by the control strategy in Fig. 2.16. In that scheme we fixed externally the flow of fresh reactant *A* feed. Also, we used reactor residence time (via the effluent flowrate) as the controlled dominant variable. Feedback information (internal reactant composition information) is provided to this controller by the ratio of the two recycle stream flows.

Step 5: Control product quality and handle safety, operational, and environmental constraints

Select the "best" valves to control each of the product-quality, safety, and environmental variables.

We want tight control of these important quantities for economic and operational reasons. Hence we should select manipulated variables such that the dynamic relationships between the controlled and manipulated variables feature small time constants and deadtimes and large steady-state gains. The former gives small closed-loop time constants and the latter prevents problems with the rangeability of the manipulated variable (control valve saturation).

It should be noted that establishing the product-quality loops first, before the material balance control structure, is a fundamental difference between our plantwide control design procedure and Buckley's procedure. Since product quality considerations have become more important in recent years, this shift in emphasis follows naturally.

The magnitudes of various flowrates also come into consideration. For example, temperature (or bottoms product purity) in a distillation column is typically controlled by manipulating steam flow to the reboiler (column boilup) and base level is controlled with bottoms product flowrate. However, in columns with a large boilup ratio and small bottoms flowrate, these loops should be reversed because boilup has a larger effect on base level than bottoms flow (Richardson rule). However, inverse response problems in some columns may occur when base level is controlled by heat input. High reflux ratios at the top of a column require similar analysis in selecting reflux or distillate to control overhead product purity.

Step 6: Fix a flow in every recycle loop and control inventories (pressures and levels)

Fix a flow in every recycle loop and then select the best manipulated variables to control inventories.

In most processes a flow controller should be present in all liquid recycle loops. This is a simple and effective way to prevent potentially large changes in recycle flows that can occur if all flows in the recycle loop are controlled by levels, as illustrated by the simple process examples in Chap. 2. Steady-state and dynamic benefits result from this flow control strategy. From a steady-state viewpoint, the plant's separation section is not forced to operate at significantly different load conditions, which could lead to turndown or flooding problems.

From a dynamic viewpoint, whenever all flows in a recycle loop are set by level controllers, wide dynamic excursions can occur in these flows because the total system inventory is not regulated. The control system is attempting to control the inventory in each individual vessel by changing the flowrate to its downstream neighbor. In a recycle loop, all level controllers see load disturbances coming from the upstream unit. This causes the flowrate disturbances to propagate around the recycle loop. Thus any disturbance that tends to increase the total inventory in the process (such as an increase in the fresh feed flowrate) will produce large increases in all flowrates around the recycle loop.

Fixing a flowrate in a recycle stream does not conflict with our discussion of picking a dominant reactor variable for production rate control in Step 4. Flow controlling a stream somewhere in all recycle loops is an important simple part of any plantwide control strategy.

Gas recycle loops are normally set at maximum circulation rate, as limited by compressor capacity, to achieve maximum yields (Douglas doctrine).

Once we have fixed a flow in each recycle loop, we then determine what valve should be used to control each inventory variable. This is the material balance step in the Buckley procedure. Inventories include all liquid levels (except for surge volume in certain liquid recycle streams) and gas pressures. An inventory variable should typically be controlled with the manipulated variable that has the largest effect on it within that unit (Richardson rule). Because we have fixed a flow in each recycle loop, our choice of available valves has been reduced for inventory control in some units. Sometimes this actually eliminates the obvious choice for inventory control for that unit. This constraint forces us to look outside the immediate vicinity of the holdup we are considering.

For example, suppose that the distillate flowrate from a distillation column is large compared to the reflux. We normally would use distillate to control level in the reflux drum. But suppose the distillate recycles back to the reactor and so we want to control its flow. What manipulator should we use to control reflux drum level? We could potentially use condenser cooling rate or reboiler heat input. Either choice would have implications on the control strategy for the column, which would ripple through the control strategy for the rest of the plant. This would lead to control schemes that would never be considered if one looked only at the unit operations in isolation.

Inventory may also be controlled with fresh reactant makeup streams as discussed in Step 4. Liquid fresh feed streams may be added to a location where level reflects the amount of that component in the process. Gas fresh feed streams may be added to a location where pressure reflects the amount of that material in the process.

Proportional-only control should be used in nonreactive level loops for cascaded units in series. Even in reactor level control, proportional control should be considered to help filter flowrate disturbances to the downstream separation system. There is nothing necessarily sacred about holding reactor level constant.

Step 7: Check component balances

Identify how chemical components enter, leave, and are generated or consumed in the process.

Component balances can often be quite subtle, but they are particularly important in processes with recycle streams because of their integrating effect. They depend upon the specific kinetics and reaction paths in the system. They often affect what variable can be used to set production rate or reaction rate in the reactor. The buildup of chemical components in recycle streams must be prevented by keeping track of chemical component inventories (reactants, products, and inerts) inside the system.

We must identify the specific mechanism or control loop to guarantee that there will be no uncontrollable buildup of any chemical component within the process (Downs drill).

What are the methods or loops to ensure that the overall component balances for all chemical species are satisfied at steady state? We can limit their intake, control their reaction, or adjust their outflow from the process.

As we noted in Chap. 2, we can characterize a plant's chemical components into reactants, products, and inerts. We don't want reactant components to leave in the product streams because of the yield loss and the desired product purity specifications. Hence we are limited to the use of two methods: consuming the reactants by reaction or adjusting their fresh feed flow. Product and inert components all must have an exit path from the system. In many systems inerts are removed by purging off a small fraction of the recycle stream. The purge rate is adjusted to control the inert composition in the recycle stream so that an economic balance is maintained between capital and operating costs.

We recommend making a Downs drill table that lists each chemical component, its input, its generation or consumption, and its output. This table should specify how the control system will detect an imbalance in chemical components and what specific action it will take if an imbalance is detected.

Step 8: Control individual unit operations

Establish the control loops necessary to operate each of the individual unit operations.

Many effective control schemes have been established over the years for individual chemical units (Shinskey, 1988). For example, a tubular reactor usually requires control of inlet temperature. High-temperature endothermic reactions typically have a control system to adjust the fuel flowrate to a furnace supplying energy to the reactor. Crystallizers require manipulation of refrigeration load to control temperature. Oxygen concentration in the stack gas from a furnace is controlled to prevent excess fuel usage. Liquid solvent feed flow to an absorber is controlled as some ratio to the gas feed. We deal with the control of various unit operations in Chaps. 4 through 7.

Step 9: Optimize economics or improve dynamic controllability

Establish the best way to use the remaining control degrees of freedom.

After satisfying all of the basic regulatory requirements, we usually have additional degrees of freedom involving control valves that have not been used and setpoints in some controllers that can be adjusted. These can be utilized either to optimize steady-state economic process performance (e.g., minimize energy, maximize selectivity) or to improve dynamic response.

For example, suppose an exothermic chemical reactor may be cooled with both jacket cooling water and brine (refrigeration) to a reflux condenser. For fast reactor temperature control, manipulation of brine is significantly better than cooling water. However, the utility cost of brine is much higher than cooling water. Hence we would like the control system to provide tight reactor temperature control while minimizing brine usage. This can be achieved with a valve position control strategy. Reactor temperature is controlled by manipulating brine. A valve position controller looks at the position of the brine control valve and slowly adjusts jacket cooling water flow to keep the brine valve approximately 10 to 20 percent open under steady-state operation (Fig. 3.2).

Additional considerations

Certain quantitative measures from linear control theory may help at various steps to assess relationships between the controlled and manipulated variables. These include steady-state process gains, open-loop time constants, singular value decomposition, condition numbers, eigenvalue analysis for stability, etc. These techniques are described in

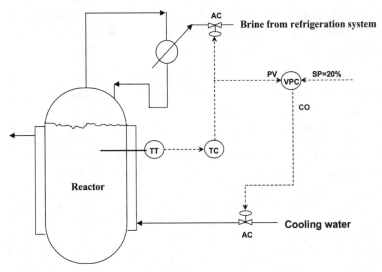

Figure 3.2 Illustration of valve position control strategy.

detail in most process control textbooks. The plantwide control strategy should ultimately be tested on a nonlinear dynamic model that captures the essential process behavior.

Since the design of a chemical process profoundly affects its dynamic controllability, another part of the problem's open-ended nature is the opportunity to change the *process* design. The design-and-control interaction problem remains as yet an open research area in terms of the plantwide control problem.

3.4   Justification of Sequence

Although the order of the steps in the design procedure may initially seem arbitrary, the sequence comes from a consideration first of choices that have already been assigned due to equipment or business constraints and then the importance in a hierarchy of priorities. Steps 1 and 2 are straightforward in determining the objectives and available degrees of freedom.

Step 3 is next because the reactor is typically the heart of an industrial process and the methods for heat removal are intrinsically part of the reactor design. So it is usually not optional what degrees of freedom can be used for exothermic reactor control. When the heat generated in an exothermic reactor is used within the process via energy integration, we must ensure that the energy is dissipated and not recycled. Hence we examine process-to-process heat exchangers and heat-integrated unit operations to determine that we have sufficient degrees of freedom (bypass lines or trim heaters/coolers).

The choice of where production rate is set (Step 4) is often a pivotal decision, but it frequently is determined externally by a business objective. This removes another degree of freedom that cannot be used. If we are free to choose the handle for production rate, then Steps 5 through 7 are the priority order. However, at Step 7 we may determine that the choice will not work in light of other plantwide control considerations, in which case we would return to Step 4 and select a different variable to set production rate. Determining the *best* choice at Step 4 can only be done via nonlinear dynamic simulation of disturbances with a complete control strategy.

Step 5 is done next because the control of product quality is closely tied to Step 1 and is a higher priority than the control of inventories. Hence it should be done early when we still have the widest choice of manipulators available. Its importance is based on the issue of variability, which we want to be as small as possible for on-aim product quality control. Variability in inventory control tends to be not as critical, which is the reason it is done in Step 6.

Only after the total process mass balance has been satisfied can we check on the individual component balances in Step 7. That then settles the plantwide issues. We now apply our knowledge of unit operation control in Step 8 to improve performance and remain consistent with the plantwide requirements. Finally, Step 9 addresses higher level concerns above the base regulatory control strategy.

This, then, is a general and straightforward method for tackling the control system design problem for an entire process. Using the procedure as a framework, we should be able to transform an initially complex and seemingly intractable problem into one that can be solved. Before we illustrate the application of the procedure to four industrial processes, we analyze and summarize the control systems for individual unit operations. We also discuss how they fit into the plantwide perspective.

3.5 Conclusion

We have discussed in detail each of the nine steps in our plantwide control design procedure. The first two steps establish the control objectives and control degrees of freedom for the plant. In the third step we discuss how the plantwide energy management problem can be converted to a local unit operation energy management problem by using the plant utility system.

The heart of the plantwide control problem lies in Steps 4 through 7, where we establish how to set production rate, maintain product quality, prevent excessive changes in recycle flowrates, control inventories, and balance chemical components. These steps demand a plant-wide perspective that often leads to control strategies differing significantly from those devised by looking at isolated unit operations.

In Part 3 we illustrate the application of these steps in four industrial processes.

3.6 References

Buckley, P. S. *Techniques of Process Control*, New York: Wiley (1964).

Cantrell, J. G., Elliott, T. R., and Luyben, W. L. "Effect of Feed Characteristics on the Controllability of Binary Distillation Columns," *Ind. Eng. Chem. Res., 34*, 3027–3036 (1995).

Douglas, J. M. *Conceptual Design of Chemical Processes*, New York: McGraw-Hill (1988).

Downs, J. J. "Distillation Control in a Plantwide Control Environment," Chap. 20 in *Practical Distillation Control*, W. L. Luyben (ed.), New York: Van Nostrand Reinhold (1992).

Luyben, M. L., Tyreus, B. D., Luyben, W. L. "Plantwide Control Design Procedure," *AIChE J., 43*, 3161–3174 (1997).

Luyben, M. L., Tyreus, B. D., Luyben, W. L. "Analysis of Control Structures for Reaction/Separation/Recycle Processes with Second-Order Reactions," *Ind. Eng. Chem. Res., 35*, 758–771 (1996).

Shinskey, F. G. *Process Control Systems,* 3d ed., New York: McGraw-Hill (1988).

Tyreus, B. D., Luyben, W. L. "Dynamic and Control of Recycle Systems: 4. Ternary Systems with One or Two Recycle Streams," *Ind. Eng. Chem. Res., 32*, 1154–1162 (1993).

Tyreus, B. D., and Luyben, W. L. "Tuning of PI Controllers for Integrator/Deadtime Processes," *Ind. Eng. Chem. Res., 31*, 2625–2628 (1992).

Control of
Individual Units

4

Reactors

4.1 Introduction

In Chap. 2 we illustrated the issues of plantwide control by using very simple unit operations. In each example, the reactor was a well-mixed, liquid-filled tank where we carried out isothermal, elementary reaction steps. A level control loop was sufficient to make the reactor fully functional. The point we tried to convey is that no matter how simple the individual unit operations (and their controls) may be, new control issues arise when the units become part of an integrated plant. Certainly these issues are still present when we introduce more complexity into the individual processing steps. In this chapter we study some industrially relevant reactor systems.

There is a vast literature on chemical kinetics and reactor engineering, but relatively little has been written on the practical aspects of industrial reactor control. Given the importance of reactors in chemical processing plants, this situation is surprising. One explanation might be that reactors are highly nonlinear so the bulk of control theory (which is for linear systems) does not readily apply. Another reason could be that many reactors are modeled as distributed systems (e.g., plug-flow reactors) and the models don't lend themselves to compact transform analysis. One can also argue that reactor control invariably involves plantwide process control, thereby significantly extending the scope of any study. Finally, many industrially important reactor systems have not been published in the open literature because they have proprietary designs and control systems.

Whatever the reasons may be for the lack of references on reactor control, we found ourselves in a difficult position in writing this chapter. The initial intent was to give a brief overview of the subject and show some typical unit operation control strategies to be used in Step 8 of our plantwide control design procedure. We would have liked to refer elsewhere regarding the details of the various control schemes. Having to abandon this approach, we concluded that a methodology toward reactor control would be a desirable substitute. A methodology has the advantage of being independent of the particular kinetic system and reactor type under consideration.

The methodology we adopt is the one developed by Reuel Shinnar and his coinvestigators. It has been described in a series of papers by Arbel et al. (1995a, 1995b, 1996, 1997). The approach revolves around four areas of activity: reactor modeling, study of open-loop reactor behavior, control structure selection, and use of process design to affect controllability. In their papers, Arbel et al. have illustrated their approach on a complex unit, a fluidized catalytic cracker. We have avoided merely reporting their results and instead tried to be additive by providing examples around much simpler systems such as CSTRs and plug-flow reactors.

The methodology on reactor control assumes some familiarity with reactor engineering. Process engineers and reactor design specialists already have mastered the fundamentals of reaction thermodynamics and kinetics but we were not sure that all control engineers felt equally comfortable with these topics. We have therefore taken the liberty of including a section on fundamentals that covers the elements of reactor engineering. The material for this section has been borrowed from the many excellent textbooks on the subject. We have found the short text by Denbigh and Turner (1971) and the more extensive work by Froment and Bischoff (1979) particularly useful. However, if this feels like old hat to you, please skip directly to the sections on models and open-loop behavior of reactors where we start the design methodology.

4.2 Thermodynamics and Kinetics Fundamentals

4.2.1 Thermodynamic constraints

Chemical reactions can occur provided the reaction products have a lower energy content than the reactants. This is analogous to a ball rolling down an incline. The ball can keep rolling as long its potential energy is lowered from the motion. The "potential" energy for a chemically reacting system, held at a constant temperature, is the Gibbs free energy G, which equals the total energy of the system minus the portion unavailable for mechanical work (see App. A for more details).

$$G = U + PV - TS = H - TS \qquad (4.1)$$

where U = internal energy
    P = pressure
    V = system volume
    T = absolute temperature
    S = entropy
    H = enthalpy

A reaction's ability to proceed at a constant temperature T hinges on a negative change in the Gibbs free energy:

$$\Delta G = \Delta H - T\Delta S \qquad (4.2)$$

where ΔG = Gibbs free energy change due to reaction
    ΔH = enthalpy change due to reaction
    ΔS = entropy change due to reaction

A system may continue to react as long as the Gibbs free energy keeps decreasing ($\Delta G < 0$). This might occur, for example, when the chemical bonds in the products are stronger than those in the reactants. Such reactions are exothermic and proceed with a heat release. An example is the formation of benzene and methane from toluene and hydrogen mentioned in Chap. 1 for the HDA process:

$$C_6H_5CH_3 + H_2 \rightarrow C_6H_6 + CH_4 \qquad (4.3)$$

$$\Delta H^0 = -42.2 \text{ kJ/mol}$$

where ΔH^0 is the heat of reaction at standard conditions (e.g., pure gaseous components at T = 298 K and P = 1 atm)

Another example discussed in more detail in this book (Chap. 11) is the formation of vinyl acetate from ethylene, acetic acid, and oxygen. This reaction is highly exothermic due to the strong and stable bonds in the water molecules.

$$C_2H_4 + CH_3COOH + \tfrac{1}{2}O_2 \rightarrow CH_2 = CHOCOCH_3 + H_2O \qquad (4.4)$$

$$\Delta H^0 = -176.4 \text{ kJ/mol}$$

While exothermic reactions are often thermodynamically favored, a negative enthalpy of reaction is not necessary. A reaction can also occur when $\Delta H > 0$ (endothermic reaction), provided that there is a large enough increase in the entropy of reaction ($\Delta S > 0$) at constant temperature. An example is the catalytic cracking of heavy petroleum fractions into gasoline, middle distillates, and light olefins. Here it takes a considerable amount of energy to break the carbon-to-carbon bonds in the heavy oil components but the energy gets distributed over a large number of small molecules, thus decreasing "order" and increasing entropy.

Since the entropy increases in most cracking operations, we might expect that it decreases for synthesis reactions. Indeed, this is often the case. A typical example is a polymerization reaction where a large number of monomer units organize in chains, thus reducing the entropy at constant temperature. Whenever a reaction is accompanied by a reduction in entropy, it must be exothermic. This follows from Eq. (4.2), since the Gibbs free energy can be negative for a decrease in entropy $(-T\Delta S > 0)$ only when the enthalpy change is negative. Since synthesis reactions are far more common than cracking operations in the chemical industry, we can understand why exothermic reactions dominate in this area.

Exothermic reactions with a decrease in entropy reach equilibrium $(\Delta G = 0)$ at some temperature and reverse beyond this point. This is evident from Eq. (4.2) where the negative term ΔH will cancel with the positive term $-T\Delta S$ when T gets sufficiently large. Since we already noted that such reactions are common in the chemical industry, should we expect most reactions to be reversible? In principle, yes, but in practice we operate many reactors at a temperature far below the equilibrium point and therefore never notice any influence of the reverse reaction. There are, however, industrially important exceptions to this rule. The manufacture of ammonia from nitrogen and hydrogen and the formation of sulfur trioxide from sulfur dioxide and oxygen are two prominent cases.

$$N_2 + 3H_2 \rightleftharpoons 2NH_3$$

$$\Delta H^0 = -91.8 \text{ kJ/mol}$$

$$\Delta S^0 = -0.198 \text{ kJ/mol} \cdot \text{K}$$

$$T_e = 91.8/0.198 = 464 \text{ K}$$

$$SO_2 + \tfrac{1}{2}O_2 \rightleftharpoons SO_3$$

$$\Delta H^0 = -98.9 \text{ kJ/mol}$$

$$\Delta S^0 = -0.094 \text{ kJ/mol} \cdot \text{K}$$

$$T_e = 98.9/0.094 = 1052 \text{ K}$$

where ΔS^0 = entropy change due to reaction at standard conditions (e.g., gases at 298 K, 1 atm)
T_e = equilibrium temperature based on standard state properties

Ammonia and sulfur trioxide reactors must be designed and operated such that the heat of reaction does not raise the reactor temperature enough to cause the reaction to reach equilibrium and therefore stop. The same applies to the reversible isomerization reaction studied in Chap. 9.

We conclude that most reaction systems in the chemical industries are exothermic. This has some immediate consequences in terms of unit operation control. For instance, the control system must ensure that the reaction heat is removed from the reactor to maintain a steady state. Failure to remove the heat of reaction would lead to an accumulation of heat within the system and raise the temperature. For reversible reactions this would cause a lack of conversion of the reactants into products and would be uneconomical. For irreversible reactions the consequences are more drastic. Due to the rapid escalation in reaction rate with temperature we will have reaction runaway leading to excessive by-product formation, catalyst deactivation, or in the worst case a complete failure of the reactor possibly leading to an environmental release, fire, or explosion.

4.2.2 Reaction rate

Thermodynamics tells us whether a chemical reaction is possible but it does not say how fast the reaction goes. For example, the Gibbs free energy favors the formation of water from hydrogen and oxygen at room temperature and atmospheric pressure ($\Delta G^0 = -228.6$ kJ/mol). However, a mixture of these gases can remain in a flask for a long time at room temperature without any noticeable water formation. The problem is that the molecules don't have enough velocity at room temperature to overcome the reaction's activation energy. It takes a spark or a catalyst to make the reaction go at any appreciable speed. When initiated with a spark the exothermic reaction generates enough heat to elevate the temperature of the entire mixture such that the reaction proceeds with explosive speed.

In general, the rate of a reaction depends upon the activation energy and temperature in an exponential fashion:

$$r = k \cdot f(C_i) \tag{4.5}$$

$$k = A_f e^{-E_a/RT} \tag{4.6}$$

where
r = specific rate of reaction (mol/L $\cdot$ s)
k = overall rate constant
$f(C_i)$ = function of concentration C_i for reacting species
A_f = pre-exponential factor
E_a = activation energy (kJ/mol)
R = universal gas constant (8.31 J/mol $\cdot$ K)

The significance of the activation energy from a control standpoint is that it dictates to what extent temperature plays a dominant role on the reaction system. To explore this relation we can evaluate the temperature derivative of Eq. (4.5):

$$\frac{dr}{dT} = f(C_i)\frac{dk}{dT} = r\,\frac{E_a}{RT^2}$$

$$\frac{dr/r}{dT} = \frac{E_a}{RT^2}$$

(4.7)

We see that the relative increase in rate per degree change in temperature is proportional to the activation energy. This means that reactions with large activation energies increase their rates more rapidly with temperature than reactions with low activation energies. For example, the activation energy for the vinyl acetate reaction [Eq. (4.4)] is 30.5 kJ/mol. Parallel to the main reaction is also a side combustion reaction consuming some of the ethylene reactant to produce carbon dioxide:

$$C_2H_4 + 3O_2 \rightarrow 2CO_2 + 2H_2O$$

(4.8)

The activation energy for the side reaction is 84.1 kJ/mol, making it nearly 3 times more responsive to temperature than the main reaction. At 453 K, where the side reaction starts playing an important role, this translates to a 5 percent increase in the side reaction per kelvin compared to less than 2 percent for the main reaction. At very high temperatures the side reaction completely dominates the picture. Control and optimization of reactor temperature is essential for economic operation.

While reactor temperature often plays a dominant role on reaction rate, it is not the only contribution to the rate expression. We also have the influence of the concentrations of the reacting species as symbolized by $f(C_i)$ in Eq. (4.5). This expression can range from simple to very complex. In the simplest form the rate of reaction is proportional to the reactant concentrations raised to their stoichiometric coefficients. This is true for an elementary step where it is assumed that the molecules have to collide to react and the frequency of collisions depends upon the number of molecules in a unit volume. In reality, matters are far more complicated. Several elementary steps with unstable intermediates are usually involved, even for the simplest overall reactions. When the intermediates are free radicals, there can be a hundred or more elementary steps. From an engineering viewpoint it is impractical to deal with scores of elementary steps and intermediates and we usually seek an overall rate expression in terms of the stable, measurable (in principle) components in the reactor. In theory we can derive such an expression from a known set of elementary steps but in practice we are well-advised to use experimental data to develop reliable rate expressions.

As an example of the often unintuitive nature of many rate expressions, we examine the vinyl acetate reaction. If this reaction took place as a series of simple elementary steps resulting from gas phase collisions of the reactants, the rate expression would be proportional to the reactant concentrations raised to their stoichiometric coefficients.

$$r_{el} = kC_E C_A C_O^{0.5} \tag{4.9}$$

$$= k(RT)^{-2.5} p_E p_A p_O^{0.5}$$

where r_{el} = rate based on "elementary" kinetics
     p_i = partial pressure of component i
     E = ethylene
     A = acetic acid
     O = oxygen

In reality the reaction is carried out over a solid catalyst and probably involves numerous elementary steps and intermediates. As a result the actual rate expression found by fitting experimental data looks quite a bit different than Eq. (4.9):

$$r_{ex} = k \frac{p_E p_A p_O (1 + 1.7 p_W)}{[1 + 0.583 p_O (1 + 1.7 p_W)](1 + 6.8 p_A)} \tag{4.10}$$

where r_{ex} = rate based on empirical expression fitted to experimental data and p_W = partial pressure of water. Notice that the partial pressure of water enters the empirical rate expression. Water is a product and does not participate as a reactant in the chemistry. We would never expect it to enter an elementary expression such as Eq. (4.9). The fact that it enters into the empirical expression must be viewed in light of the reaction occurring over a catalyst. It is conceivable that water somehow promotes the catalyst activity.

From a control standpoint it is important to know the actual form of the kinetic rate expression. The reason is that it dictates which reactant and product components play a dominant role on the behavior of the reactor. For example, components with a nonzero exponent play a direct and predictable role especially at low concentrations. But even components that don't enter the rate expression (e.g., inerts) play an indirect role by their influence on the concentrations of the reactants. This secondary effect is much less predictable for a multicomponent system. In general, it is seldom easy to determine the composition effects of a single component from a superficial look at the reaction and its stoichiometry. Instead, the best available kinetic rate equation should be incorporated into a reactor model (steady-state or dynamic). With this model we then can evaluate the integrated effects of temperature and species compositions. For example, from the overall stoichiometry of the vinyl acetate reaction we would expect that an increase in the partial pressure of acetic acid should increase the total production rate. This turns out to be incorrect because the rate is close to zero order in acetic acid partial pressure due to cancellation effects with the denominator in Eq. (4.10). Acetic acid can therefore not be counted as a dominant variable for the control of a vinyl acetate reactor with the kinetic expression derived from the given experimental data.

In most rate expressions the dominant components are reactants. Occasionally we encounter reactions that are dominated by a product. Such reactions are said to be *autocatalytic*. A simple example of an autocatalytic scheme is

$$A + 2B \rightarrow 3B + C$$

with a rate expression

$$r = kC_B^2 C_A \tag{4.11}$$

Since we make one extra mole of component B for every two moles of B consumed, and the rate is highly dependent on the concentration of B, the product dominates the reactor behavior especially at low concentrations.

4.2.3 Multiple reactions

In most industrially relevant reacting systems, one main reaction typically makes the desired products and several side reactions make byproducts. The specific rate of production or consumption of a particular component in such a reaction set depends upon the stoichiometry and the rates. For example, assume that the main reaction for making vinyl acetate, Eq. (4.4), proceeds with a rate r_1 (mol/L $\cdot$ s) and that the side reaction, Eq. (4.8), proceeds with rate r_2 (mol/L $\cdot$ s). Then the net consumption of ethylene is $(-1)r_1 + (-1)r_2$ (mol/L $\cdot$ s). Similarly, the net consumption of oxygen is $(-0.5)r_1 + (-3)r_2$, and the net production of water is $(1)r_1 + (2)r_2$. For a given chemistry (stoichiometry), our ability to control the production or consumption of any one component in the reactor is thus limited to how well we can influence the various rates. This boils down to manipulating the reactor temperature and/or the concentrations of the dominant components. Occasionally, the reaction volume for liquid-phase reactions or the pressure for gas-phase reactions can also be manipulated for overall production control. These are the fundamentals of reactor control.

4.2.4 Conversion, yield, and selectivity

While the individual reaction rates are the variables that can be affected in a reacting system, we often express the performance of the reactor in terms of measures derived from the rates. Conversion and yield are such quantities. Conversion refers to the fractional consumption of a reactant in the reactor feed, whereas yield refers to the amount of product made relative to the amount of a key reactant fed to the reactor. In recycle systems the per-pass conversion of the various reactants is a relevant measure. It depends upon the rate of reaction for the specific component but also on the reactor feed. The per-pass conversion of an excess reactant is less than that of a limiting reactant. For example, the per-pass conversion of ethylene in a typical vinyl acetate reactor is only 7 percent whereas the per-pass conversion of oxygen is 36 percent. In Chap. 2 we discussed the plantwide control implications of incomplete conversion.

Selectivity is a measure similar to yield in that it tells how much desirable product is made from a reactant. Selectivity is particularly informative for parallel reactions involving the same key reactant. Here it measures what proportion of the converted reactant goes to useful products. In the production of vinyl acetate, for example, ethylene reacts in the main reaction as well as in the side reaction. The selectivity of ethylene to vinyl acetate is

$$\text{SEL} = 100 \, \frac{\text{mol/s vinyl acetate produced}}{\text{mol/s vinyl acetate produced} + 0.5 \text{ mol/s } CO_2 \text{ produced}}$$

4.3 Fundamentals of Reactors

4.3.1 Types

The simplest type of reactor we can imagine is an adiabatic batch reactor. In such a system we charge the reactants, close the vessel, and let the mixture react for a certain amount of time. The progress of such a reaction depends upon the initial conditions of the batch and the kinetic rate expressions. For a sufficiently long batch time relative to the reaction rates, the system ends up at a unique steady state. This steady state is a thermodynamic equilibrium state due to the Gibbs free energy reaching a minimum (isothermal reactor) or the entropy reaching a maximum (adiabatic operation).

While the adiabatic batch reactor is important and presents many control issues in its own right, we are concerned here primarily with continuous systems. We consider in detail two distinct reactor types: the continuous stirred tank reactor (CSTR) and the plug-flow reactor. They differ fundamentally in the way the reactants and the products

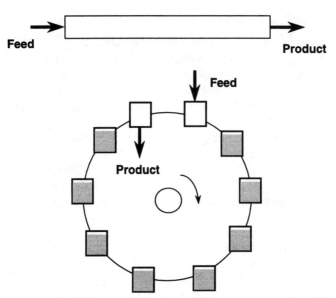

Figure 4.1 Plug-flow reactor represented as sequence of small batch reactors.

are backmixed within the reactor. The plug-flow reactor shown schematically in Fig. 4.1 has no backmixing between compartments of the reactor. An ideal plug-flow reactor can be visualized as a "conveyor system" of small adiabatic batch reactors that are charged with fresh feed, allowed to react in isolation, and then discharged at the end of the cycle (Fig. 4.1). Each little reactor in the system will have its own composition and temperature depending only upon the initial conditions (the fresh feeds), the batch time, the thermodynamics, and the kinetics of the system.

Instead of discharging all the material from each little batch reactor in Fig. 4.1, imagine now that we leave some material in the reactor to be mixed with the fresh feed (Fig. 4.2). Diluting the fresh feeds will alter the reaction rate. In addition, we must run the conveyor system faster to make up for the reduced space in the reactors. These two factors change the reactor output such that it is no longer simply determined by the fresh feed conditions and the batch time but depends upon the recycle ratio as well. We can, however, eliminate the explicit mention of the recycle ratio from the system description by recycling everything (complete backmixing) and running the conveyor system infinitely fast. All the small batch reactors now have the same composition and temperature. This is a simple analog of a CSTR where the backmixing is typically done by agitation, sparging, or fluidization as in Fig. 4.3.

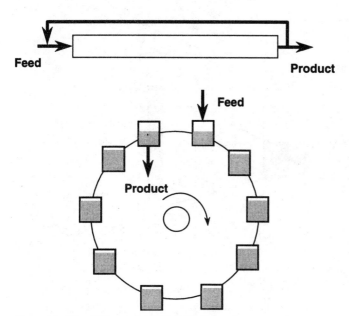

Figure 4.2   Plug-flow reactor with recycle.

The CSTR and the tubular reactor are idealized extremes on the mixing spectrum. Industrial reactors usually fall somewhere in between. For example, diffusion, conduction, convection, and flow channeling can make a plug-flow reactor have less than ideal composition and temperature profiles. Similarly, insufficient agitation in a CSTR can create channeling that alters its homogeneous composition and temperature assumption. Some industrial reactors can be viewed as hybrids between a CSTR and a plug-flow reactor. For example, a fluidized bed can be considered backmixed in the solid phase whereas the plug-flow assumption is frequently valid for the gas phase. The same

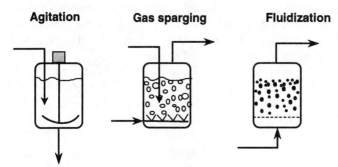

Figure 4.3   Continuous stirred tank reactors.

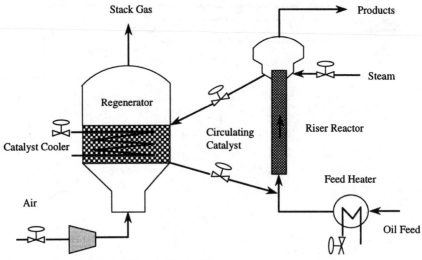

Figure 4.4 Fluidized catalytic cracking unit.

can be said about sparged gas-liquid reactors. More complex reactor systems can be constructed by combining plug-flow and backmixed units. For example, in Fig. 4.4 we show a schematic of a fluidized catalytic cracker that can be viewed as a plug-flow reactor (the riser reactor) coupled to a CSTR (the regenerator).

4.3.2 Reactor selection

The primary reason for choosing a particular reactor type is the influence of mixing on the reaction rates. Since the rates affect conversion, yield, and selectivity we can select a reactor that optimizes the steady-state economics of the process. For example, the plug-flow reactor has a smaller volume than the CSTR for the same production rate under isothermal conditions and kinetics dominated by the reactant concentrations. The opposite may be true for adiabatic operation or autocatalytic reactions. For those situations, the CSTR would have the smaller volume since it could operate at the exit conditions of a plug-flow reactor and thus achieve a higher overall rate of reaction.

While reactor size may be important for the economics of the process, other factors such as yield and selectivity typically play a greater role. A classic situation is when the main products suffer degradation by consecutive reactions. In the scheme

$$A \to B \to C \tag{4.12}$$

we are interested in the product, B, and would like to minimize the by-product, C. The appropriate measure of reactor performance is the yield of B. It is not difficult to visualize that a plug-flow reactor typically gives the highest yield for these systems. First, the average concentration of A is high in the plug-flow reactor, thus promoting the formation of B. Second, the concentration of B gradually builds up in the plug-flow reactor and can be made to reach its maximum at the exit. In contrast, the CSTR has a low concentration of A and a high concentration of B throughout the reactor. These factors slow the production of B and promote the formation of C. The plug-flow reactor is therefore the best choice. See Denbigh and Turner (1971) for further discussions on the chemical factors affecting the choice of reactor.

As is often the case, a completely different set of factors influences the choice of reactor from a control standpoint. Our main focus for control is stability and responsiveness to changes in the manipulated variables. We look at the details of these issues in the following sections but make some broad brush generalizations at this point.

The key to controllability is simplicity. From that standpoint the adiabatic, plug-flow reactor is hard to beat. As we have mentioned earlier, its performance is a unique function of the feed conditions. For isothermal operation of exothermic reactions, the CSTR is first in line. The reason is that there is only one temperature and one set of compositions to control as opposed to the profiles that govern tubular reactors. In addition, the cooled plug-flow reactor can be very sensitive to operating parameters. Beyond this it is hard to generalize. This is why we feel that a methodology toward reactor control is the best approach, as we described in the introduction to this chapter. This methodology begins by characterizing the open-loop behavior of reactors.

4.4 Models

4.4.1 Introduction

Chemical reactors are inherently nonlinear in character. This is primarily due to the exponential relationship between reaction rate and temperature but can also stem from nonlinear rate expressions such as Eqs. (4.10) and (4.11). One implication of this nonlinearity for control is the change in process gain with operating conditions. A control loop tuned for one set of conditions can easily go unstable at another operating point. Related to this phenomenon is the possibility of *open-loop instability* and *multiple steady states* that can exist when there is material and/or thermal recycle in the reactor. It is essential for the control engineer to understand the implications of nonlinearities and what can be done about them from a control standpoint as well as from a process design standpoint.

To study issues related to nonlinearities we need nonlinear models. In principle there is no difficulty in finding all the equations that can go into a model for an arbitrary reactor. Textbooks on reaction engineering do a great job in covering the basics as well as the details. The trick is to produce a practical model in the sense that it is computationally efficient and yet provides the required insights about the system. This is unfortunately still an art. In this respect we refer to two excellent papers by Shinnar (1978, 1981). Other useful references to dynamic modeling are Luyben (1990) and Arbel et al. (1995a).

We will take a closer look at one of the simplest systems conceivable, a constant-volume and -density, cooled CSTR with a first-order, irreversible reaction $A \rightarrow B$. While this model is quite simple it still contains most of the relevant issues surrounding an open-loop, nonlinear reactor. Referring to Fig. 4.5, this system can be described by one component balance and one energy balance:

$$V \frac{dC_A}{dt} = F(C_{A0} - C_A) - rV \qquad (4.13)$$

$$(V\rho C_P + M_w C_w) \frac{dT}{dt}$$
$$= F\rho (C_{P0}T_0 - C_P T) - UA(T - T_c) + rV(-\Delta H) - \Delta Q \quad (4.14)$$

where V = reaction volume
F = volumetric flow through reactor
C_{A0}, C_A = concentrations of component A in fresh feed and product
r = rate of reaction
ρ = density
C_{P0}, C_P = heat capacity of feed and product streams
M_w = mass of metal wall in reactor
C_w = heat capacity of metal wall in reactor
T_0, T, T_c = feed, reactor, and jacket temperatures
UA = heat transfer capacity of cooling jacket
ΔH = heat of reaction
ΔQ = reactor heat loss

This model formulation assumes that the reactor wall temperature on the coolant side is the same as the temperature of the reactor contents. The rate of reaction is modeled as

$$r = kC_A = A_f e^{-E_a/RT} C_A \qquad (4.15)$$

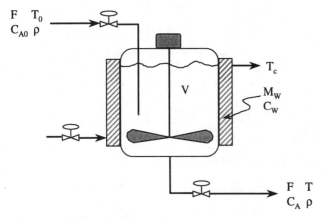

Figure 4.5 CSTR model nomenclature.

Equations (4.13) to (4.15) represent a nonlinear, dynamic reactor model suitable for simulation on a digital computer. From this model we can derive other models that are useful for control explorations of the reactor.

4.4.2 Nonlinear steady-state model

A nonlinear steady-state model is obtained by setting the derivatives equal to zero in Eqs. (4.13) and (4.14). This gives a set of nonlinear algebraic equations that normally have to be solved numerically. However, in this particular case we can find an explicit solution for C_A in terms of temperature.

$$C_A = \frac{C_{A0}}{1 + \tau A_f e^{-E_a/RT}} \qquad (4.16)$$

where $\tau = V/F$ is the residence time of the reactor.

Equation (4.16) is combined with Eqs. (4.15) and (4.14) to produce the following implicit equation in temperature:

$$\frac{\tau A_f e^{-E_a/RT}}{1 + \tau A_f e^{-E_a/RT}} \frac{C_{A0}}{\rho} (-\Delta H)$$

$$= \left(C_P + \frac{UA}{F\rho}\right) T - C_{P0}T_0 + \frac{\Delta Q - UAT_c}{F\rho} \qquad (4.17)$$

Equations (4.16) and (4.17) are required to explore issues around output multiplicity and steady-state sensitivity.

4.4.3 Linear dynamic model

The usefulness of linear dynamic models is familiar to all control engineers. Linear models can be written compactly in matrix notation.

$$\dot{\mathbf{x}} = \mathbf{A}\mathbf{x} + \mathbf{B}\mathbf{u} \tag{4.18}$$

where $\mathbf{x}$ = vector of state variables
$\dot{\mathbf{x}}$ = vector of time derivatives
$\mathbf{u}$ = vector of inputs
$\mathbf{A}$ and $\mathbf{B}$ = constant coefficient matrices

Since we presently are concerned only with open-loop issues, we will focus our attention on the $\mathbf{A}$ matrix.

A linear model can be derived from its nonlinear counterpart by linearization. We rearrange the original set of nonlinear differential equations [Eqs. (4.13) to (4.14)] such that the time derivatives are explicit functions of all the states.

$$f_1 = \frac{dC_A}{dt} = \frac{(C_{A0} - C_A)}{\tau} - r \tag{4.19}$$

$$f_2 = \frac{dT}{dt} = \frac{F\rho(C_{P0}T_0 - C_PT) - UA(T - T_c) + rV(-\Delta H) - \Delta Q}{V\rho C_P + M_w C_w} \tag{4.20}$$

The derivatives are then expanded as Taylor series around a steady state. For example, the first few terms of f_1 take the following form:

$$f_1 = \bar{f}_1 + \left(\frac{\partial f_1}{\partial C_A}\right)_T (C_A - \bar{C}_A) + \left(\frac{\partial f_1}{\partial T}\right)_{C_A} (T - \bar{T}) + \cdots \tag{4.21}$$

where the overbar means that the variable is evaluated at steady-state conditions. When we truncate the Taylor series after the linear terms and introduce perturbation variables, $\tilde{C}_A$ and $\tilde{T}$, defined as

$$\tilde{C}_A \equiv C_A - \bar{C}_A$$

$$\tilde{T} \equiv T - \bar{T}$$

we obtain the desired result

$$\begin{pmatrix} \dfrac{d\tilde{C}_A}{dt} \\[2ex] \dfrac{d\tilde{T}}{dt} \end{pmatrix} = \begin{bmatrix} \dfrac{\partial f_1}{\partial C_A} & \dfrac{\partial f_1}{\partial T} \\[2ex] \dfrac{\partial f_2}{\partial C_A} & \dfrac{\partial f_2}{\partial T} \end{bmatrix} \begin{pmatrix} \tilde{C}_A \\[2ex] \tilde{T} \end{pmatrix} \tag{4.22}$$

The **A** matrix, which is the Jacobian of the original equation set, has a compact analytical form for the simple CSTR:

$$\mathbf{A} = \begin{bmatrix} -\dfrac{1 + \overline{k}\tau}{\tau} & -\dfrac{\overline{k}\,\overline{C}_A E_a}{R\overline{T}^2} \\[3ex] \dfrac{V(-\Delta H)\overline{k}}{V\rho C_P + M_w C_w} & \dfrac{-F\rho C_P - UA + \overline{k}\,\overline{C}_A V(-\Delta H)E_a/R\overline{T}^2}{V\rho C_P + M_w C_w} \end{bmatrix} \quad (4.23)$$

For more complicated systems the Jacobian must be evaluated numerically.

4.5 Open-Loop Behavior

4.5.1 Multiplicity and open-loop instability

No matter how well we design a plant's control system, there will be times when the plant operators feel they need to intervene by switching some controllers into manual and running the process open-loop. When they do that, they certainly expect the process to respond to changes in the valve positions. They also expect that each unit operation will find a unique steady state for a given set of valve loadings. But what if the process can produce different results for the same constant inputs? Or what if the process can start oscillating or run away while the control valves are held in constant positions? This would be most confusing and undesirable. These phenomena, related to output multiplicity and open-loop instability, can occur in chemical reactors if not properly considered during reactor design.

The idea that a unit operation could have two or more steady states for the same values of the input variables is not only confusing in practice but somewhat hard to understand conceptually. We will try to explain the situation, first in words and then graphically. The verbal explanation of multiplicity centers around two of the necessary conditions: nonlinearity and process feedback.

The need for nonlinearity is easy to see. A linear equation has no more than a single solution. A quadratic equation may have two solutions, etc. The describing equations for a reactor must therefore be nonlinear to show output multiplicity.

We next turn to process feedback. We mentioned earlier that a plug-flow reactor can be viewed as a string of small batch reactors. We also pointed out that the result of each batch is uniquely determined by the fresh feeds since the solution to the batch equations is a forward integration in time. A plug-flow reactor cannot by itself show output multiplicity or open-loop instability. This picture changes when we start recycling material from the end of one batch to the beginning of the next. Now, the output is no longer uniquely determined by the fresh feeds but depends, in a sense, on its own value. The ambiguity created by the recycle (process feedback) makes it possible to have more than one solution for the same fixed fresh feed conditions. Since CSTRs have infinite internal recycle ratios they are perfect candidates for output multiplicity and open-loop instabilities.

The graphic illustration of output multiplicity focuses on the steady-state solutions of Eq. (4.17). This equation can be viewed as a trade-off between a nonlinear heat generation term, $Q_{gen}(T)$, and a linear heat removal expression:

$$Q_{rem}(T) = a_1 T - a_0$$

In Fig. 4.6 we have plotted a typical heat generation expression (curve a) along with the heat removal line, b. In this case the two curves intersect at three locations corresponding to three different reactor conditions that are possible for the same operating parameters and feed conditions. The low-temperature steady state is uneconomical since the feeds are virtually unconverted. The highest-temperature steady state has nearly complete conversion but may be too hot. Under those conditions side reactions may set in or the reactor pressure becomes too high. The middle steady state strikes a good compromise and is where

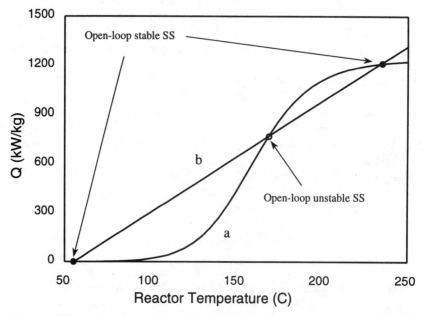

Figure 4.6 Heat generation and heat removal curves for CSTR.

we would like to operate. Will the reactor stay there when the operator switches the reactor temperature controller into manual? The answer is no! The middle steady state is always open-loop unstable when there are three steady states. It is easy to see why. At the middle steady state the heat generation curve has a steeper slope than the heat removal line causing the reactor to drift away from this point. A small increase in temperature, for example, will stimulate more heat production than heat removal and the temperature goes up further yet until the reactor reaches the hottest steady state. Therefore, operation at an open-loop unstable steady state, even if it is feasible with feedback control, is undesirable and should be avoided by proper design of the reactor and its heat exchanger. We refer to the original papers by van Heerden (1953, 1958) and to Arbel et al. (1995b) for further discussions on output multiplicity and unstable steady states in reactors with process feedback.

4.5.2 Open-loop oscillations

How can we change the design of the reactor and its cooling system so that the middle steady-state in Fig. 4.6 is stable? Based on the reason why the middle steady state is unstable we need to increase the slope of the heat removal line beyond that of the heat production curve. This is the so-called slope condition. A quick glance at Eq. (4.17) reveals that the slope condition could be satisfied by providing more heat transfer area in the cooler (increase UA). Fortunately this is not hard to do at the design stage. Figure 4.7 shows a design where the slope condition is met such that there is only one intersection between the curves. Surely this single steady-state must be stable? Not necessarily. The slope condition is a necessary condition for stability but it is not sufficient. A complete analysis involves examining the dynamic stability from the linearized system as shown by Bilous and Amundson (1955). They pointed out that the roots γ_1, γ_2 of the open-loop characteristic equation for a two-dimensional system have the following values

$$\gamma_1, \gamma_2 = \frac{-(a_{11} + a_{22}) \pm \sqrt{(a_{11} + a_{22})^2 - 4(a_{11}a_{22} - a_{12}a_{21})}}{2} \quad (4.24)$$

where $(-a_{ij})$ is the element in row i, column j of the Jacobian matrix $\mathbf{A}$ and γ_i are the eigenvalues of $\mathbf{A}$. For a two dimensional system it is clear that the roots will have negative real parts if, and only if,

$$a_{11} + a_{22} > 0$$

$$a_{11}a_{22} - a_{12}a_{21} > 0$$

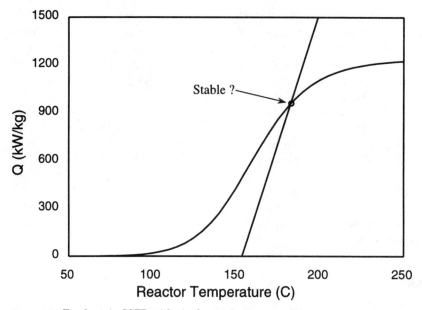

Figure 4.7 Exothermic CSTR with single steady-state condition.

To demonstrate the importance of performing an eigenvalue analysis on a reactor system with backmixing we cite an industrial example reported in a paper by Vleeschhouwer et al. (1992). A schematic of their 6000-L oxidation reactor with heat exchange is shown in Fig. 4.8. The authors claim that a simple CSTR model provides a good representation of the actual system. From the information provided in their paper we plotted the heat generation and removal curves to verify that the slope condition was met (see Fig. 4.7). We next inserted the reactor and operating parameters into Eqs. (4.23) and (4.24) to arrive at the following eigenvalues for the steady state given in the paper.

$$\gamma_1, \gamma_2 = \frac{0.00024 \pm i\sqrt{9.1 \times 10^{-5}}}{2} \qquad (4.25)$$

The steady state is evidently dynamically unstable since the eigenvalues have positive real parts. In addition, the eigenvalues are complex, indicating that the system will move away from its unstable steady state in an oscillatory fashion.

Figure 4.9 shows the results of a dynamic simulation we performed featuring the open-loop behavior of a backmixed reactor that satisfies the slope condition for steady-state stability but has dynamically unstable roots. Table 4.1 contains the reactor parameters and operating conditions used in the model, as listed by Vleeschhouwer et al. (1992).

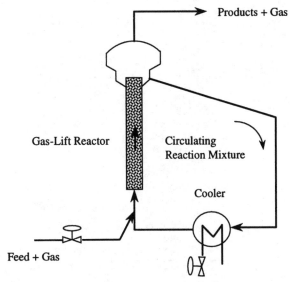

Figure 4.8 Oxidation reactor.

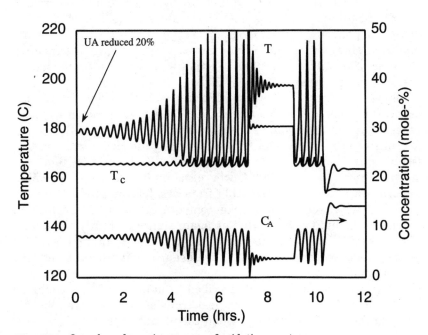

Figure 4.9 Open-loop dynamic response of oxidation reactor.

TABLE 4.1   Reactor Operating Parameters for
Open-Loop Oscillations

Heat capacity of feed	2700 J/kg·K
Heat capacity of product	2400 J/kg·K
Density	650 kg/m³
Feed temperature	303 K
Normalized activation energy (E_a/R)	11,000 K
Heat of reaction	−159,000 J/mol
Mass of metal wall in reactor	10,000 kg
Heat capacity of metal wall	460 J/kg·K
Heat losses	175,000 W
Steady-state temperature	182°C
Coolant temperature	163 → 170°C
Feed flowrate	3.2 kg/s
Feed concentration	7.81 mol/kg
Pre-exponential factor	9.7×10^7 s⁻¹
Heat transfer parameter (UA)	98,000 W/K
Residence time	17.7 min
Effective liquid volume	5.2 m³

We show the reactor and jacket temperatures (T and T_c) along with C_A, the concentration of component A in the reactor. Initially, when the simulation started, the heat transfer area was sufficiently large to maintain open-loop static and dynamic stability. However, a few minutes into the simulation, we reduced UA by 20 percent. This creates dynamic instability with complex eigenvalues as in Eq. (4.25). The reactor temperature and composition start oscillating with a growing amplitude. However, the amplitude growth stops roughly 5 hours after the onset of instability and the reaction enters into a *limit cycle* of constant period and amplitude.

This type of limit cycle was actually observed in the industrial reactor described by Vleeschhouwer et al. (1992). It was also shown by these authors that stability could be regained by altering the operating point. We confirmed this in our simulation by raising the jacket temperature after roughly 7 hours of process time (see Fig. 4.9). The reactor temperature transitions to nearly 200°C where the limit cycle stops. The high temperature may not be desirable from a safety or product quality standpoint, so we reduced the jacket temperature after the reactor had stabilized. The limit cycle started up again almost instantly. We finally reduced the jacket temperature to about 155°C. The oscillations stopped but the exit concentration might have been too high for the refining train to handle.

The operational difficulties described around the CSTR could be avoided altogether by making the heat transfer area in Fig. 4.8 sufficiently large. This is easy to do when the plant is being designed but may be a costly proposition once the process is in operation. That is why it is so important to perform adequate controllability studies before the plant is built.

4.5.3 Parametric sensitivity

Open-loop multiplicity, instability, and oscillations are restricted to reactors with material or energy feedback. It is tempting then to think that plug-flow reactors, lacking backmixing, should be problem-free. This is far from the case. Ironically, it is the lack of backmixing that causes some of the problems. For example, while a CSTR can operate adiabatically at a reasonable temperature and conversion given a low feed temperature, a plug-flow reactor requires a minimum feed temperature to get the reactions started at the inlet of the reactor. In addition, from this point on every additional unit of heat produced along the adiabatic plug-flow reactor is locally added to what has already been produced upstream. There is no temperature averaging over the length of the reactor. This introduces a climbing temperature profile as long as there are additional reactants to convert. Since the rate of reaction increases exponentially with temperature, there is always a possibility that the reaction will go to completion, which may lead to unacceptably high temperatures. The adiabatic temperature rise ΔT_{ad} and the maximum reactor temperature $T_{\max}$ are important parameters for plug-flow systems. The adiabatic temperature rise can be estimated from the reaction heat available in the feed in relation to its capacity to absorb this heat. The maximum temperature is then the feed temperature plus the adiabatic temperature rise:

$$T_{\max} = T_0 + \Delta T_{ad} \approx T_0 + \frac{y_{A0}(-\Delta H)}{C_{P0}^m} \qquad (4.26)$$

where T_0 = feed temperature
   y_{A0} = mole fraction of limiting reactant in feed
   ΔH = heat of reaction
   C_{P0}^m = average molar heat capacity of feed

We illustrate the use of Eq. (4.26) by calculating the maximum achievable temperature in the HDA and vinyl acetate reactors discussed in this book. Both of these are gas-phase, plug-flow systems.

Starting with the HDA reactor, we find most of the needed information in Chap. 10. The feed temperature is 1150°F, the heat of reaction $-21,500$ Btu/lb · mol, and the mole fraction of toluene (limiting component) in the reactor feed is 0.0856. The molar heat capacity of the feed is computed from its composition and standard literature data:

$$C_{P0}^m = 0.43(7.2) + 0.48(16) + 0.09(59) = 16 \text{ Btu/lb} \cdot \text{mol} \cdot {}^\circ\text{F}$$

The estimated maximum temperature is

$$T_{max} \approx 1150 + \frac{(0.0856)(21{,}500)}{16} = 1265°F = 685°C$$

The adiabatic temperature rise for this system is roughly $\Delta T_{ad} =$ 115°F = 64°C.

The vinyl acetate process is described in Chap. 11. The reactor inlet temperature is 148.5°C and oxygen is the limiting reactant ($y_{A0} = 0.075$). The heat of reaction is -42.1 kcal/mol vinyl acetate or -84.2 kcal/mol oxygen. The average heat capacity of the feed is computed from data provided in Chap. 11:

$$C_{P0}^m = 0.59(13.3) + 0.22(14.2) + 0.11(37.5) + 0.08(7.5)$$

$$= 15.7 \text{ kcal/kmol } °C$$

When we consider the main reaction and exclude the side reaction, the estimated maximum temperature is

$$T_{max} \approx 148.5 + \frac{(0.075)(84{,}200)}{15.7} = 551°C$$

and the adiabatic temperature rise, $\Delta T_{ad} = 402°C$.

The maximum temperature in the HDA reactor is greater than it is in the vinyl acetate reactor. Yet, only the HDA reactor can be operated adiabatically; the vinyl acetate reactor must be cooled. There are two reasons for this. The first reason is the magnitude of the adiabatic temperature rise. There is virtually no reaction system that stays selective over a 400°C temperature range. For example, the side reaction in the vinyl acetate reactor becomes significant over 200°C and dominates completely at 500°C. In addition, the catalyst in the vinyl acetate reactor would quickly deactivate over 200°C due to sintering. The HDA reactor, on the other hand, has no catalyst and the adiabatic temperature rise is under 100°C, which is a reasonable range for a reaction to remain selective.

The second reason why the vinyl acetate reactor must be cooled is sensitivity. Sensitivity S is a measure of the reaction's potential to run away from the feed temperature. This tendency is determined by two factors: the relative increase in reaction rate with temperature [Eq. (4.7)] and the feed's potential to elevate the reactor temperature, ΔT_{ad}:

$$S = \left(\frac{dr/r}{dT}\right)_{T_0} \Delta T_{ad} = \frac{E_a}{RT_0^2} \cdot \frac{y_{A0}(-\Delta H)}{C_{P0}^m}$$

The sensitivity for the HDA reactor is 10.8, which means that the reactor exit stream could react at a rate roughly 11 times faster than the feed provided the compositions remained constant throughout the reactor. This kind of amplification in rate due to temperature indicates that there is a good chance the reaction will go to completion, thereby inducing the full adiabatic temperature rise.

The vinyl acetate reaction is even more sensitive. With $S = 44.2$ for the side reaction alone it would be virtually impossible to try to prevent an adiabatic reactor from reaching full conversion on oxygen by mere control of the inlet temperature. Small changes in the inlet conditions would quickly amplify down the reactor, forcing complete conversion, adiabatic temperature rise, and destruction of the catalyst, if we were lucky in the best scenario. We must apply external cooling to the vinyl acetate reactor.

Given that the vinyl acetate reactor has to be cooled, we now examine the issues around cooled plug-flow systems in general. These systems often attain their maximum temperature before the exit, resulting in a temperature peak or hot spot somewhere along the length of the reactor (see Fig. 4.10). The height of the temperature peak is designed to be well under the adiabatic temperature rise. However, the adiabatic temperature rise is always lurking in the background and cooled reactors can also be quite sensitive to minor changes in the system's operating parameters. These issues were first discussed by Bilous and Amundson (1956) and were later investigated in detail by Froment and Bischoff (1979). A sensitive reactor can easily run away (but also quench) due to minor changes in the feed conditions or the coolant temperature as shown in Fig. 4.10. We emphasize that the runaway and quenched states are not caused by output multiplicity but are distinct steady states related to different, but narrowly separated, input conditions.

From a control perspective, it is important to know if a plug-flow reac-

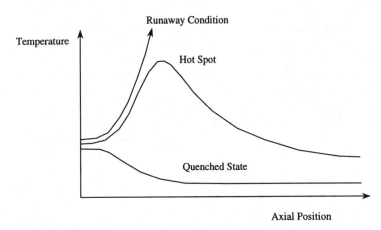

Figure 4.10 Plug-flow reactor temperature sensitivity.

tor has a hot spot that is sensitive to the operating parameters. If this is the case it might be necessary to measure and control the hot-spot temperature directly to prevent runaway. It is of course best to avoid the dynamic control problem altogether and have an insensitive reactor by design. Froment and Bischoff (1979) and Shinnar et al. (1992) discuss various design criteria for making a plug-flow reactor insensitive to parameter variations. In principle it is a matter of providing enough cooling surface per unit reactor volume in relation to the rate of heat generation per unit volume. The required ratio of surface area to heat generated increases as a function of the system sensitivity S.

The vinyl acetate reactor we use in Chap. 11 has been designed to be insensitive to parameter variations under normal operating conditions. The hot-spot temperature is only 162°C with an exit temperature of 159°C. It is adequate to control the exit temperature instead of the hot spot. Since multiplicity, open-loop stability, and sensitivity are of no concern for this reactor, we can focus our attention on the open-loop characteristics relevant to the control of exit temperature with jacket cooling.

Of prime interest for any control loop is the linearity of the process for varying operating conditions. For example, as the catalyst deactivates we have to operate at a higher exit temperature to maintain vinyl acetate production rate. A steady-state model is useful to explore what happens to the process under such conditions. The model helps us make two important observations about the reactor as the exit temperature increases. First, the hot spot moves out of the reactor and the highest temperature occurs at the exit. The reason is that the side reaction, with its high activation energy and high heat of reaction, plays an increasingly greater role at higher temperatures. The second change in reactor character is an increase in process gain. The gain is measured as the ratio of the change in exit temperature to the change in coolant temperature. Figure 4.11 shows the reactor gain as a function of the exit temperature and with the reactor feed temperature as a parameter. The large change in gain will impact how we tune the temperature controller. Gain scheduling may be necessary.

We can explain why the gain changes with temperature by looking at the log mean temperature driving force and the heat transferred to the coolant. The following equation is valid only when the reactor inlet and exit temperatures are higher than the coolant temperature and the reactor temperature increases linearly. This is approximately true for the vinyl acetate reactor.

$$Q_c = UA \, \frac{\Delta T_{\text{out}} - \Delta T_{\text{in}}}{\ln \dfrac{\Delta T_{\text{out}}}{\Delta T_{\text{in}}}}$$

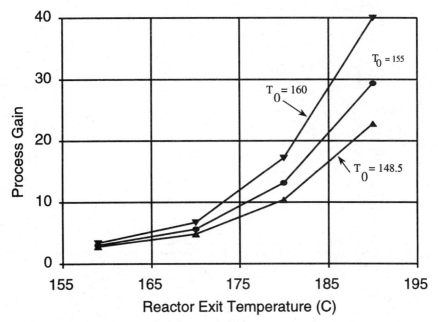

Figure 4.11 Vinyl acetate reactor gain as function of exit temperature.

where Q_c = heat transferred to coolant
$\Delta T_{out} = T - T_c$
$\Delta T_{in} = T_0 - T_c$
T = reactor exit temperature
T_0 = reactor feed temperature
T_c = coolant temperature

A decrease in ΔT_{in} (due to an increase in the coolant temperature) leads to a correspondingly larger increase in ΔT_{out} for the same heat transferred. However, the heat generated by the reaction increases exponentially with temperature, forcing the exit temperature to climb further. This relation amplifies with large differences between ΔT_{out} and ΔT_{in} as well as with large, average activation energies.

4.5.4 Wrong-way behavior

The final open-loop reactor issue we discuss is the problem of inverse response or *wrong-way behavior* as it is called in the reactor engineering literature. The inverse response refers to the temporary increase in the exit temperature in some packed, plug-flow reactors following a decrease in the feed temperature (Fig. 4.12). The wrong-way behavior stems from the difference in propagation speed between concentration

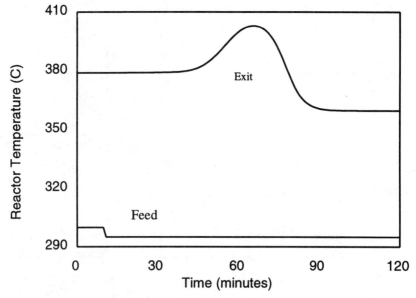

Figure 4.12   Wrong-way effect in packed plug-flow reactor.

and temperature disturbances. For example, when the feed temperature decreases, the conversion in the front end of the reactor decreases.
When the higher concentration of reactants reaches the middle and
back end of the reactor, where the packing is still hot, the reaction
takes off and produces a transient temperature increase. Eventually
the lower feed temperature will produce a new steady state with a
lower exit temperature than before.

Inverse response creates control difficulties. Assume, for example,
that we wish to control the exit temperature of an adiabatic plug-flow
reactor by manipulating the inlet temperature as shown in Fig. 4.13.
From a steady-state viewpoint this is a perfectly reasonable thing to
consider, since there are no issues of output multiplicity or open-loop
instability, assuming the fluid is in perfect plug flow and there is no

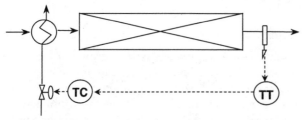

Figure 4.13   Proposed temperature control of adiabatic plug-
flow reactor.

conduction in the packing. We also assume that the main reactant is nearly consumed at the exit. This means that the reactor cannot show parametric sensitivity because there is no more reactant left to increase the temperature in the reactor. However, as all control engineers know, the dynamic problem of inverse response severely limits the closed-loop performance achievable by any type of controller. For a stand-alone reactor the control problem can be avoided by controlling the reactor inlet instead of the exit temperature. However, when the reactor is part of an integrated process with feed preheating, this solution may not be sufficient. It is important to understand what design factors promote wrong-way behavior so that the phenomenon can be properly considered at the design stage.

The effect of various design parameters on wrong-way behavior in reactors has been thoroughly investigated by Luss and coworkers (Metha et al., 1981; Pinjala et al., 1988; Il'in and Luss, 1992 and 1993). Pinjala et al. (1988) used a simple model that is quite instructive in showing the influence of some of the design factors on the severity of wrong-way behavior. The model is written for a first-order reaction, $A \rightarrow B$, carried out in a packed tubular reactor. The model does not assume perfect plug flow since it allows for dispersion of material and energy. A single component balance and the energy balance take the following forms when written in dimensionless variables.

$$\frac{1}{\text{Le}} \frac{\partial x}{\partial t} = \frac{1}{\text{Pe}_m} \frac{\partial^2 x}{\partial z^2} - \frac{\partial x}{\partial z} - \text{Da} \cdot e^{-(1/y)} \cdot x$$

$$\frac{\partial y}{\partial t} = \frac{1}{\text{Pe}_h} \frac{\partial^2 y}{\partial z^2} - \frac{\partial y}{\partial z} + \beta \cdot \text{Da} \cdot e^{-(1/y)} \cdot x + U(y_w - y)$$

where x = dimensionless composition variable, C_A/C_{A0}
$\quad\quad y$ = dimensionless temperature, RT/E_a
$\quad\quad t$ = dimensionless time, $(ut'/\epsilon L)/\text{Le}$
$\quad\quad z$ = dimensionless axial position, z'/L

The other parameter groups in the model are defined as follows:

Lewis number (ratio of thermal time constant to material time constant due to fluid flow):

$$\text{Le} = 1 + \frac{(1 - \epsilon)\rho_s C_{Ps}}{\epsilon \rho_0 C_{P0}}$$

Peclet number for mass (ratio of linear flowrate to diffusivity):

$$\text{Pe}_m = \frac{Lu}{D_e}$$

Peclet number for heat (ratio of fluid heat flow to conductivity):

$$\text{Pe}_h = \frac{Lu\rho_0 C_{P0}}{k_e}$$

Damköhler number (ratio of residence time to reaction time):

$$\text{Da} = \frac{L}{u}A_f$$

Normalized adiabatic temperature rise (ratio of available reaction heat to feed heat capacity normalized by the activation energy):

$$\beta = \frac{R}{E_a}\frac{(-\Delta H)C_{A0}}{\rho_0 C_{P0}}$$

Dimensionless heat transfer parameter:

$$U = \frac{2hL}{ru\rho_0 C_{P0}}$$

where D_e = effective dispersion coefficient for mass
      h = overall heat transfer coefficient
      k_e = effective thermal conductivity
      L = reactor length
      r = radius of reactor
      t' = time
      u = superficial velocity
      z' = axial position coordinate
      ϵ = void fraction of bed
      ρ_0 = feed density
      ρ_s = packing density
      C_{P0} = heat capacity of feed
      C_{Ps} = heat capacity of packing

   The Lewis number has been defined by Ray and Hastings (1980) as the ratio of two time constants, τ_h and τ_c. The thermal time constant τ_h is the ratio of the total heat capacity of a unit reactor volume V to the heat capacity of the fluids flowing through this volume. For an unpacked reactor this time constant equals the residence time, $\tau = V/F$. For a packed reactor the thermal time constant can be considerably longer than the residence time. Similarly, the material time constant τ_c is the ratio of material holdup in a unit reactor volume to the flow of material through this volume. Again, for an unpacked reactor, $\tau_c = \tau$, whereas a packed reactor could have the material time constant much shorter than the overall holdup time due to the space occupied by the packing. Pinjala et al. (1988) point out that only packed reactors can exhibit wrong-way behavior. They also show that the maximum temperature peak is independent of the Lewis number when Le > 100.

As an example of how to compute the Lewis number we use the vinyl acetate reactor described in Chap. 11. The catalyst bulk density, $(1 - \epsilon)\rho_s$, is given as 385 g/L and the heat capacity of the packing, C_{Ps}, is 0.23 cal/g · °C. We previously calculated the fluid heat capacity to be 15.7 cal/g · mol · °C. The feed molar density is

$$\frac{P}{RT} = \frac{(128/14.7)}{0.08205(273.15 + 148.5)} = 0.25 \text{ g} \cdot \text{mol/L}$$

Porosity data is given for the catalyst as $\epsilon = 0.8$. The Lewis number for the vinyl acetate reactor is

$$\text{Le} = 1 + \frac{(1 - \epsilon)\rho_s C_{Ps}}{\epsilon \rho_0 C_{P0}} = 1 + \frac{385(0.23)}{0.8(0.25)(15.7)} = 29$$

This means that it takes the fluid stream 29 times longer to cause a temperature change in the reactor than it takes to change the reactor composition. This is a sufficient difference in the propagation speeds to induce wrong-way behavior. However, a large Lewis number is not sufficient to create a problem. Another important requirement is that the reaction should be near completion at the exit of the reactor. This requirement is not met for the vinyl acetate reaction which has only a 36 percent conversion in oxygen.

For a reactor system with a significant Lewis number and a reaction that is near completion at the reactor exit, Pinjala et al. (1988) showed that the maximum peak of the temperature transient y^* is an increasing function of $\text{Pe}_h \beta / \text{Da}$:

$$y^* \propto \frac{\text{Pe}_h \beta}{\text{Da}} \tag{4.27}$$

Equation (4.27) predicts that the temperature peak will be severe when the adiabatic temperature rise is large and when there is little thermal conductivity in the system ($\text{Pe}_h \gg 1$). By allowing the reactor to have significant dispersion of heat (through conduction or backmixing) it is possible to reduce the height of the temperature peak. However, as we allow for dispersion we introduce thermal feedback and create the potential for output multiplicity and instability. For example, it is possible that the entire reactor can switch to a second hot steady state following a decrease in feed temperature. This phenomenon starts as a temperature wave, initiated by the wrong-way behavior, moving slowly backward up the reactor until it reaches the feed point. Here, the feed mixture "ignites" and the entire reactor quickly switches to the high-temperature steady state. This ignition can have a disastrous impact on the packing or the integrity of the reactor.

4.6   Unit Control

4.6.1   Heat management

We noted earlier in this chapter that many reactions in the chemical industries are exothermic and require heat removal. A simple way of meeting this objective is to design an adiabatic reactor. The reaction heat is then automatically exported with the hot exit stream. No control system is required, making this a preferred way of designing the process. However, adiabatic operation may not always be feasible. In plug-flow systems the exit temperature may be too hot due to a minimum inlet temperature and the adiabatic temperature rise. Systems with backmixing suffer from other problems in that they face the awkward possibilities of multiplicity and open-loop instability. The net result is that we need external cooling on many industrial reactors. This also carries with it a control system to ensure that the correct amount of heat is removed at all times.

The control system must manipulate heat removal from the reactor, but what should be the measured (and controlled) variable? Temperature is a good choice because it is easy to measure and it has a close thermodynamic relation to heat. For a CSTR, temperature control is particularly attractive since there is only one temperature to consider and it is directly related to the heat content of the reactor. However, in a spatially distributed system like a plug-flow reactor the choice of measured variable is not so clear. A single temperature is hardly a unique reflection of the excess heat content in the reactor. We may select a temperature where the heat effects have the most impact on the operation. This could be the hot spot or the exit temperature depending upon the design of the reactor and its normal operating profile.

To summarize, we find that heat management is required for many industrial reactors and that this task can be accomplished by controlling a reactor temperature. We start our discussion of unit control by reviewing several reactor temperature control schemes.

Continuous stirred tank reactors.   The simplest method of cooling a CSTR is shown in Fig. 4.14. Here we measure the reactor temperature and manipulate the flow of cooling water to the jacket. Using a jacket for cooling has two advantages. First, it minimizes the risk of leaks and thereby cross contamination between the cooling system and the pro-

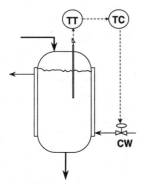

Figure 4.14 Simple CSTR temperature control.

cess. Second, there are no internals to obstruct an agitator from providing effective mixing. The main disadvantage of using jacketed cooling is the limitation on heat transfer area due to reactor geometry.

Direct supply of cooling water is a simple, reliable, and inexpensive method for cooling the reactor. However, it has some serious shortcomings. First, the water flow at low rates may be inadequate to maintain a good heat transfer coefficient. Second, the gain between the coolant flow and the amount of heat transferred varies nonlinearly with load, thereby making controller tuning difficult. Third, there can be a significant temperature gradient on the cooling water side that could create local hot spots on the process side.

Fortunately we can readily solve many of the problems associated with a direct supply of cooling water. For example, in Fig. 4.15 we have provided a water recirculation loop to maintain a large constant flow of water through the jacket. Fresh cooling water is added to the loop to maintain the desired reactor temperature. This arrangement keeps

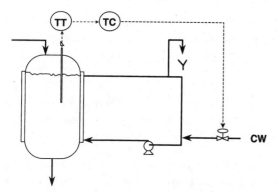

Figure 4.15 Circulating cooling water temperature control.

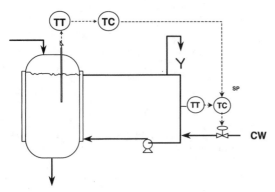

Figure 4.16 Cascade control.

the heat transfer coefficient constant, it linearizes the process gain, and it maintains a uniform temperature in the jacket. The only potential drawback with the arrangement in Fig. 4.15 is the dynamics added by the cooling water loop. Backmixing in the jacket loop averages out changes made in the cooling water supply. The entire jacket loop has to change temperature before there is a change in heat transfer from the reactor. This temperature lag is in series with the thermocouple lag and the thermal lag of the reactor content. With a low controller gain the reactor temperature could change slowly, causing sluggish control. With a high controller gain the jacket temperature may overshoot the correct value and cause the reactor temperature loop to be underdamped. The solution is to provide a cascade arrangement as shown in Fig. 4.16. The secondary controller maintains a target jacket temperature and the reactor temperature controller provides the setpoint. The jacket temperature controller can be tuned tightly, which effectively shortens the secondary lag caused by the cooling jacket dynamics. This makes it possible to tune the reactor temperature controller for better performance.

It is particularly important to keep the lags small when controlling a CSTR around an open-loop unstable point. This is easy to visualize in the root locus diagram shown in Fig. 4.17. Here the process transfer function is third order with one unstable pole stemming from an open-loop unstable steady state (the middle steady state in Fig. 4.6). The other two time constants come from the thermowell and the cooling jacket dynamics. Figure 4.17 shows that a minimum controller gain K_{min} is required to stabilize the reactor with a proportional-only controller. However, when the gain becomes too large ($K_c > K_{max}$) the loop goes unstable again. We can increase K_{max} by moving the stable poles further to the left in the diagram. This is possible only by keeping the lags small.

The scheme shown in Fig. 4.16 can be extended to high-temperature

Third-order openloop unstable:

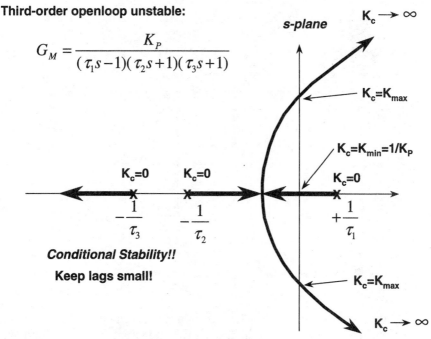

$$G_M = \frac{K_P}{(\tau_1 s - 1)(\tau_2 s + 1)(\tau_3 s + 1)}$$

Figure 4.17   Root-locus plot of open-loop unstable CSTR.

applications. For example, in Fig. 4.18 we have closed the circulation loop thus allowing for higher pressure on the jacket side or a different coolant than water. The external heat exchanger still uses cooling water. To obtain faster temperature control and to avoid nonlinearities in response to cooling water flow, we provide a bypass around the external exchanger. The cooling water flow can now be kept constant while the bypass provides a rapid response in jacket temperature.

Another method for dealing with high reactor temperatures is to generate steam, as shown in Fig. 4.19. Here we allow the coolant to boil and thereby provide a constant jacket temperature. The secondary loop controls pressure in the boiler drum by venting steam. Fresh boiler feed water is added by level control. A potential problem with this arrangement is the possibility for boiler swell that results in an increase in the level due to increased vaporization in the jacket. The increased level due to swell reduces the intake of boiler feed water when in reality it should be increased. This problem can be overcome by providing a ratio controller between the steam flow and the feed water with the ratio reset by the steam drum level controller. Boiler feed water flow will now change in the correct direction in response to load.

The heat removal schemes we have examined so far all use a cooling

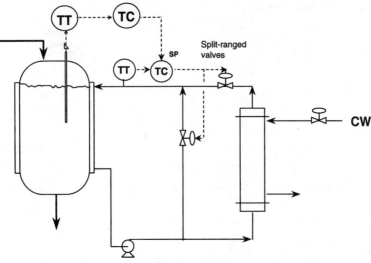

Figure 4.18 Use of external coolant heat exchanger for temperature control.

jacket. The main limitation with this approach is the amount of surface area we can make available for a given reactor volume. When we examine the open-loop characteristics of the reactor we may find that we need a larger surface area to avoid multiplicity and open-loop oscillations. Let us examine a few control approaches for systems with a large heat transfer area.

The first approach is to insert cooling coils in the reactor as shown in Fig. 4.20. The advantage with this method is that we don't have to handle the reactor content outside the reactor. The heat transfer is

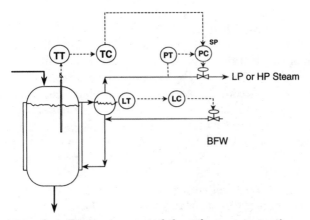

Figure 4.19 Temperature control through steam generation.

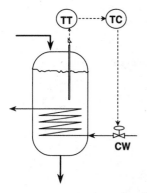

Figure 4.20 Extending heat transfer area through use of cooling coils.

also direct with no significant secondary lags. The disadvantage is that the coils can develop tube leaks that can cause contamination of the cooling water or put water in the reactor. Another problem with internal coils is that they make mechanical agitation more difficult. It may not be possible to put all the required surface area inside the reactor and still have room for an effective agitation device.

The second approach to enlarge the cooling surface is shown in Fig. 4.21. Here we bring the fluid reactor contents outside the reactor to cool the stream in an external heat exchanger. The industrial reactor shown in Fig. 4.8 used this principle. Also some fluidized-bed reactors, where the excess recycled gaseous reactants provide fluidization, commonly have these designs. There are two advantages with this method. First, the heat transfer area is completely independent of reactor size and geometry. Second, the circulation helps promote effective backmix-

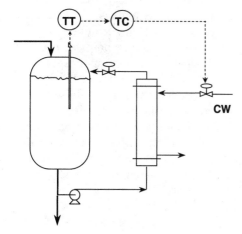

Figure 4.21 Circulation of reactor content through external heat exchanger.

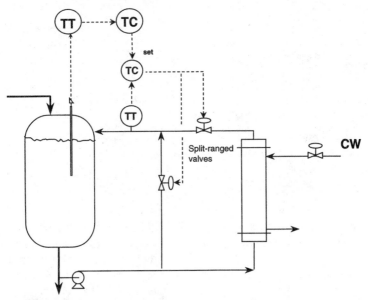

Figure 4.22 Bypass of circulating reactor content around external heat exchanger.

ing in the reactor. The main disadvantage is that we must be able to pump the reaction mass for liquid-filled reactors or compress the gases for fluidized beds.

The external heat exchanger shown in Fig. 4.21 introduces an undesirable and potentially troublesome secondary lag. To reduce the time constant due to the capacitance of the metal in the external exchanger, we can keep the coolant flow constant and bypass the exchanger as shown in Fig. 4.22. The secondary loop controls the mixed stream temperature coming from the exchanger and the bypass. This loop can be made fast, which reduces the negative effects of secondary lags.

Liquid-filled CSTRs operating close to the boiling point of a major component or a solvent offer another effective means for temperature control. The method shown in Fig. 4.23 takes advantage of the heat of vaporization to cool the reactor. The vapors enter a condenser that provides cold reflux back to the reactor. A pressure controller operates on the condenser cooling valve. The reactor temperature controller provides the pressure setpoint. While this method offers nearly perfect temperature control (due to self-regulation) under normal operating conditions, it also has some pitfalls. First, we must design the gravity return system properly. There must be a U-leg seal to set up the proper pressure balance for correct flow in the condenser. The condenser elevation must also be high enough to handle the maximum reflux load without having liquid back up in the condenser and reduce the heat

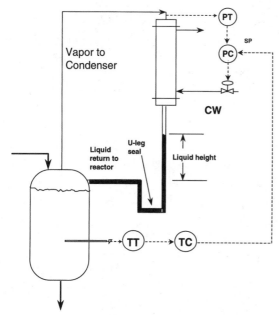

Figure 4.23 Cooling through vaporization of reactor content.

transfer surface. Second, we must provide a vent valve to remove non-condensables that could accumulate and blanket the condenser. Finally, it may be advantageous to provide a jacket with cooling water to help bring down the reactor temperature after a shutdown when the reaction has stopped and there is little vaporization.

In some situations the dynamics of the cooling system may be such that effective temperature control cannot be accomplished by manipulation of the coolant side. This could be the situation for fluidized beds using air coolers to cool the recirculating gases or for jacketed CSTRs with thick reactor walls. The solution to this problem is to balance the rate of heat generation with the net rate of removal by adjusting a reactant concentration or the catalyst flow. Such a scheme is shown in Fig. 4.24.

Plug-flow reactors. Many of the techniques discussed for cooling jacketed CSTRs can also be used for plug-flow systems. This works particularly well when the plug-flow reactor is implemented as a tube-and-shell heat exchanger. In such configurations the reaction takes place in the tubes (frequently packed with catalyst) and the coolant is on the shell side. However, plug-flow systems introduce some new issues beyond those for jacketed CSTRs. First, which temperature should we control? The exit temperature would be most convenient for a multitubular reactor since it avoids having to insert thermocouples inside

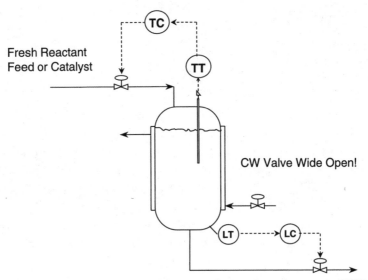

Figure 4.24 Temperature control by manipulation of reaction rate in CSTR.

narrow tubes with catalyst. On the other hand, if the hot spot is sensitive to operating conditions, we may have to control the peak temperature in the reactor. Since the hot spot frequently moves, depending upon the state of the catalyst and other operating parameters, it may be necessary to provide several measurements along the length of the reactor. The peak temperature can then be selected by using a high selector as shown in Fig. 4.25.

The second issue for cooled tubular reactors is how to introduce the coolant. One option is to provide a large flowrate of nearly constant temperature, as in a recirculation loop for a jacketed CSTR. Another option is to use a moderate coolant flowrate in countercurrent operation as in a regular heat exchanger. A third choice is to introduce the coolant cocurrently with the reacting fluids (Borio et al., 1989). This option has some definite benefits for control as shown by Bucalá et al. (1992). One of the reasons cocurrent flow is advantageous is that it does not introduce thermal feedback through the coolant. It is always good to avoid positive feedback since it creates nonmonotonic exit temperature responses and the possibility for open-loop unstable steady states.

The vinyl acetate reactor discussed in Chap. 11 uses boiling water on the shell side of the reactor. This arrangement is similar to that shown in Fig. 4.19.

It is sometimes necessary to achieve better control of the reactor temperature profile than can be accomplished in a cooled multitubular arrangement. For those cases we can arrange a series of short packed-

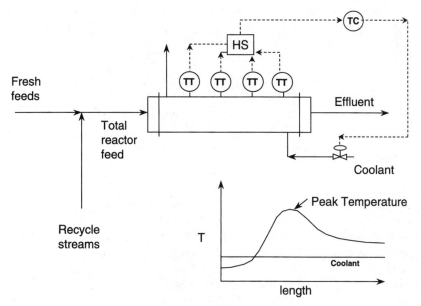

Figure 4.25   Peak temperature control in tubular reactor.

bed reactors with intermediate cooling (see Fig. 4.26). Each bed is operated adiabatically with an inlet temperature low enough to prevent complete conversion. This arrangement is also useful for reversible, exothermic reactions where the adiabatic temperature rise causes the reaction to reach equilibrium. Conversion stops at equilibrium and does not resume until the reaction mixture has been cooled in the intermediate heat exchangers. Ammonia and sulfur trioxide reactors are operated in this fashion.

Figure 4.27 shows another method of controlling plug-flow systems. Instead of cooling the effluents from each adiabatic step in a heat exchanger, we introduce a cold shot of fresh feeds. The cold shot technique increases the concentration of reactant in all the segments. Mixing the cold feed with the reactor effluent lowers the inlet temperature to the next reactor.

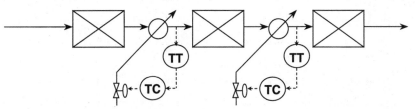

Figure 4.26   Intermediate cooling in sequence of packed adiabatic reactors.

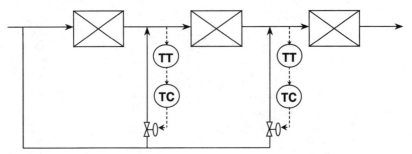

Figure 4.27 Cold-shot cooling.

A similar technique is applied to low-density polyethylene reactors. Some of these systems operate in cooled tubular reactors at a very high pressure. Since the reactor has a thick tube wall, the temperature response to changes in the coolant is slow. Instead, the reaction rate (and thereby temperature) is controlled by injecting initiator at select places along the length of the reactor tube (see Fig. 4.28).

4.6.2 Economic control objectives

Management of the heat removal in chemical reactors is required to meet thermodynamic constraints. In a sense, this is no different than satisfying mass, energy, or component balances. However, what makes heat management especially important is the close connection between heat and temperature. Temperature affects the reaction rates, which in turn control the rate of heat generation. Furthermore, in reversible reaction systems high temperature limits conversion. Proper heat management (through temperature control) becomes essential for safe, stable, and economic operation.

While safe operation is a necessary requirement to run a process it

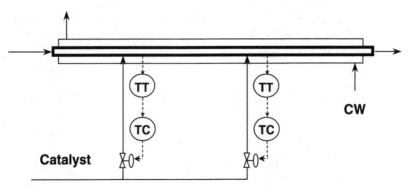

Figure 4.28 Temperature control by manipulation of reaction rate in tubular reactor.

is typically insufficient to meet all the economic objectives. These might include throughput, conversion, yield, selectivity, composition, molecular weight, density, viscosity, and color.

To control the economic objectives we must have measurements and manipulated variables. However, in the example reactors we have looked at so far it should be clear that we have only a limited number of manipulated variables, especially after we have taken care of the heat management issues. How is it then possible to achieve any level of economic control of a reactor? The answer lies in a concept introduced by Shinnar (1981) called *partial control*. In short it means that only a few dominant variables in the process (e.g., temperatures, key components) are identified, measured, and controlled by feedback controllers. Then, by varying the setpoints for the dominant variables, it becomes possible to position the process such that all the important economic variables stay within acceptable ranges. We will elaborate more on this important concept in the next section but first we introduce the classification of reactor variables used by Shinnar.

Figure 4.29 shows a block diagram of a reactor with manipulated inputs U, other measured inputs W, and unknown or unmeasured inputs N. We may assume that this reactor is more complicated than a simple plug-flow reactor or a CSTR. It may be more along the lines of the fluidized catalytic cracker that we showed in Fig. 4.4. The reactor can be described by a set of nonlinear differential equations as we have previously demonstrated. This results in a set of dynamic state variables X. The state vector is often of high dimension and we normally only measure a subset of all the states. Y is the vector of all measurements made on the system.

For the discussions around partial control it is convenient to classify the measurements further. For control purposes we pay particular attention to the variables Y_d (d for dynamic) that are measured continu-

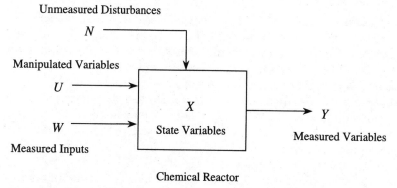

Figure 4.29 Reactor block diagram.

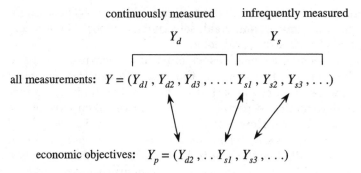

continuously measured   infrequently measured

$$Y_d \qquad\qquad Y_s$$

all measurements: $Y = (Y_{d1}, Y_{d2}, Y_{d3}, \ldots Y_{s1}, Y_{s2}, Y_{s3}, \ldots)$

economic objectives: $Y_p = (Y_{d2}, \ldots Y_{s1}, Y_{s3}, \ldots)$

Figure 4.30 Classification of measured variables.

ously or with a sample time that is small compared to the process time constant. Examples are temperature, pressure, level, density, and many on-line analyzer measurements. The rest of the measured variables are denoted Y_s (s for slow, sampled, or steady-state), such that $Y_s + Y_d = Y$. Examples of members belonging to Y_s are infrequent laboratory measurements and yield and conversion calculations based on mass balances. The list of economic variables or specifications is given the symbol Y_p (p for product or property). These variables form another subset of Y where we borrow some variables from Y_d and some variables from Y_s. Figure 4.30 summarizes the classification of measured variables.

We can make a similar classification around U and W. For instance, control valves belong to the set U_d, whereas the regeneration of a packed-bed catalyst would be classified as U_s. Similarly, measurements of the reactor feed flow and temperature belong to W_d, while a once-per-shift analysis of the reactor feed composition belongs to W_s.

4.6.3 Partial control

Control engineers know that it takes one manipulated variable for each measured variable we wish to control to setpoint. When the number of controlled variables equals the number of manipulated variables we pair up the different variables and use PI controllers for regulation. Sometimes we are fortunate to have more manipulated variables than control specifications. We can then optimize the use of the manipulators while controlling to setpoint (e.g., valve position control). Sometimes, however, the number of control objectives exceeds the number of available manipulators and we cannot control all variables to setpoint. This is when the concept of partial control is useful.

There are two premises behind partial control. The first is that many of the economic objectives are correlated. This means that by controlling some of the measured variables we can bring the rest into an acceptable range of compliance. For example, in a reaction $A \to B \to C$ we can usually improve the yield of B by limiting the conversion of A. Since yield and conversion are economic objectives we can satisfy both within a range by controlling only one variable. The reason this principle works, we believe, is that the state variables in process systems (especially reactors) are intimately linked through stoichiometry, kinetics, thermodynamics, and mixing.

The second premise behind partial control is the concept of dominance. We introduced this concept earlier when we discussed reaction rate expressions. There we mentioned that temperature often plays a dominant role for the rate of reaction especially when the activation energy is high. We also mentioned that a key component in the rate expressions can dominate the rate particularly when the component has a low concentration (e.g., limiting reactant or the catalyst).

If we accept these two premises, the implementation of partial control involves two conceptually simple steps:

1. Control all dominant variables to setpoint with feedback controllers using manipulators with a rapid response. This ensures *unit control*.

2. Adjust the setpoints of the controlled variables such that all economic objectives are brought within acceptable ranges. This is partial control.

In symbols we can express partial control as follows:

$$Y_{p,\min} < Y_p < Y_{p,\max} \tag{4.28}$$

$$Y_p = M^s(U_s, Y_{cd}^{\text{set}}, W) \tag{4.29}$$

where $M^s(\cdot)$ stands for a steady-state, correlation model.

The economic objectives Y_p can be positioned in the desired range when there is a strong relationship between the economic objectives and the measured inputs W, the steady-state manipulators U_s, and the setpoints of the controlled, dynamic variables Y_{cd}^{set}.

A few comments about the method are warranted. The controlled (dominant) variables, Y_{cd}, should be measured such that they belong to the set Y_d for rapid control. Similarly, the manipulators in the feedback control loops should belong to the set, U_d. The feedback controllers should have integral action (PI controllers). These can be tuned with minimal information (e.g., ultimate gain and frequency from a relay test). The model M^s is usually quite simple and can be developed from operating data using statistical regressions. This works because the model includes all the dominant variables of the system, Y_{cd}, as independent variables by way of their setpoints, Y_{cd}^{set}. The definition of dominance is that the whole state vector is strongly correlated to these few variables.

On the surface it might appear that partial control does not require a first-principles model for its implementation. After all, M^s is a regression model and controller tuning is based on relay-feedback information. For simple systems this may be correct. However, for most industrially relevant systems it is not intuitively obvious what constitutes the dominant variables in the system and how to identify appropriate manipulators to control the dominant variables. This requires nonlinear, first-principles models. The models are run off-line and need only contain enough information to predict the correct trends and relations in the system. The purpose is not to predict outputs from inputs precisely and accurately, but to identify dominant variables and their relations to possible manipulators.

Let's look at some examples. First consider the vinyl acetate reactor discussed in Chap. 11. It is a plug-flow system with external cooling. To satisfy the heat balance we have already proposed to close one loop around the reactor, namely between the reactor exit temperature and the coolant temperature (steam pressure). This provides us with one setpoint, Y_{cd}^{set}, that we can use to meet economic objectives, Y_p, provided exit temperature is a dominant variable.

Before we investigate the dominance aspects of the vinyl acetate reactor we ask "What are the economic objectives?" As a minimum we would like to control the production rate of vinyl acetate, T/P, and the selectivity SEL to vinyl acetate. Therefore, any dominant variable should have a significant impact on these economic objectives. Figure 4.31 shows how the objectives vary with reactor exit temperature. It is clear that we have identified a dominant variable and that it is possible to set the reactor exit temperature such that both the production rate and the selectivity fall within certain ranges for given values of the feed conditions, W, and a given level of catalyst activity, U_s. This is the meaning of partial control.

To complete the picture we should investigate if there are more dominant variables in the system. We have already touched on this issue in the section on reaction rates. There we noted that the acetic acid concentration to the reactor is not dominant. We can also argue that the ethylene partial pressure is not likely to be a dominant variable since ethylene enters the reactor in large excess. However, oxygen is the limiting component and it plays a role in the main reaction as well as in the side reaction. Oxygen therefore affects the economic objectives and is considered dominant. Feedback control of the oxygen concentration to the reactor is necessary if we want complete control of the unit.

The next example is the HDA reactor presented in Chap. 10. It is an unpacked gas flow reactor operated adiabatically so the reactor does not have any heat management control loops associated with it. First,

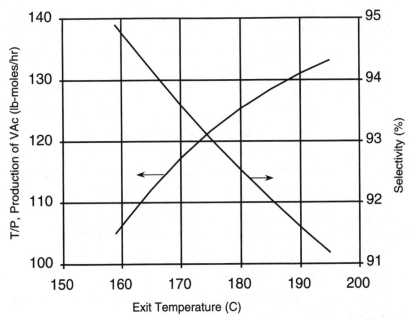

Figure 4.31   Production rate and selectivity as functions of vinyl acetate reactor temperature.

we examine the economic objectives. We would like to set the production rate. We are also interested in the yield of benzene from toluene. We use yield instead of selectivity since it is the product, benzene, that forms the by-product, diphenyl.

It is easy to imagine that inlet temperature might be a dominant variable for exothermic, adiabatic plug-flow reactors when the reactions have reasonably high activation energies. First, these systems require a minimum inlet temperature to get the reactions going at all. Second, the exit temperature depends directly upon the inlet temperature through the adiabatic temperature rise. Last, the response to inlet temperature should be unique since there are no issues of multiplicity or parametric sensitivity for the reactor in isolation. Therefore, to close a loop around $Y_{cd} = T_0$ we need only identify the appropriate manipulated variable. This variable should be able to alter the heat content of the feed stream as shown in Fig. 4.32. Since the HDA reactor requires a furnace for its operation, the choice is not difficult; we use the heat input to the furnace to control the reactor inlet temperature. The only complications we might experience is when the furnace is made small due to a very high level of heat integration in the plant. Inlet temperature control may suffer under those circumstances. We shall have more to say about this in Chap. 5.

The HDA reaction is also dominated by the partial pressure of toluene

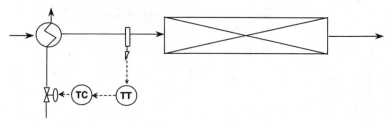

Figure 4.32 Recommended temperature control of adiabatic plug-flow reactor.

since this is the limiting component. Closed-loop control of the inlet temperature and the concentration of toluene would provide unit control. When only the inlet temperature is controlled we have partial control of the unit. This may not present a problem as long as the controlled dominant variable is among the most dominant in the system. However, when the most dominant variable is uncontrolled we are likely to encounter difficulties. The final example on dominance provides an illustration.

This example involves a hypothetical, liquid phase reaction with the following two steps:

$$A + 2B \rightarrow 3B + C \qquad r_1 = k_1 C_B^2 C_A$$

$$B \rightarrow D \qquad r_2 = k_2 C_B$$

Note that the first reaction is autocatalytic in component B. Reactant A is the limiting component. Component B is the desired product while C and D are by-products. Both reactions have moderate heats of reaction and relatively low activation energies. To maximize the yield of B, a plug-flow reactor was selected. The economic objectives are throughput and yield. What is the appropriate partial control scheme for this reactor?

Following the arguments around the HDA reactor, we conclude that inlet temperature should be a dominant variable. However, in this case it is not strongly dominant due to the low activation energies. In other words, k_1 and k_2 do not vary much with temperature. Instead, the most dominant variable is C_B, the concentration of product along the reactor. The component B enters the reactor at a low concentration and dominates the rate of both reactions at least until most of the reactant A is consumed. Furthermore, the main reaction is autocatalytic in B such that the rate of formation of B depends upon it own concentration. Unit control therefore requires that we find a manipulated variable that allows us to vary the inlet concentration of B to the reactor. Without such a loop the output from the reactor would depend entirely on minute variations in the feed concentration, which would be unpredictable. To

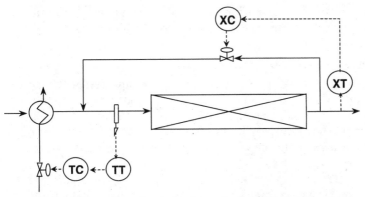

Figure 4.33 Addition of recycle stream for control of autocatalytic reaction.

gain control we have to make a design change to the process. One possibility is to create a small recycle stream from the exit of the reactor to the feed (see Fig. 4.33). This would provide the required degree of freedom to adjust the inlet concentration of B for proper unit control. The drawback of the recycle stream from a steady-state standpoint is that we give the recycled product B a chance to convert into undesirable by-product D. The steady-state yield will decline but the reactor is under unit control. This is another example of the trade-off between steady-state economics and controllability that we frequently encounter.

4.7   Design and Control

4.7.1   Process design versus controller design

As we explained in Chap. 1, process and control engineers have traditionally had distinct and different roles in designing a plant. Process engineers are present from project start and are responsible for flowsheet development and equipment design, whereas control engineers enter later and work with the designed process to specify sensors, valves, and the control system. We believe that process engineers impact controllability far more than control engineers can do through controller design. In other words, a small change to the process (such as providing 20 percent more surface area for cooling or providing a small recycle stream for unit control) can make the difference between an unstable reactor and a well-behaved one. Similarly, no amount of Kalman filtering, model predictive control, or nonlinear control theory can make up for a missing control degree of freedom. We now give some specific recommendations on how to apply process design to improve process controllability for reactor systems.

4.7.2 Design for simplicity

We pointed out earlier in this chapter that simplicity is the key to controllability. What we mean by simplicity is that the list of dominant variables should be kept as short as possible. It is easy to construct reactors where the number of dominant variables far exceed the number of manipulated variables or where the dominant variables interact in a complicated way.

Chemical reactions are irreversible processes. They occur because there is a composition gradient in the reactor. However, as the reactions proceed they can create other gradients such as pressure profiles, mixing gradients, and temperature profiles. The thermodynamic variables affected by the various gradients also influence the chemical kinetics so that the whole system can be coupled in a complex way. This coupling, or feedback, promotes the chances for multiplicity and open-loop instability. The more dominant variables that are affected and the more of those that interact, the less are our chances of controlling and modeling the system. This is particularly true when the stoichiometry and kinetics are complex as well. In fact, we would like to offer the following recommendation based on our own experiences but nicely formulated by Shinnar (1997): *Simple reactions can be carried out in complex reactors, but complex reactions need simple reactors.*

So, what kind of complications should be avoided? The most prominent ones are uncertain hydrodynamics, mixing, and mass transfer. These phenomena are difficult to predict by themselves and when they interact and affect the rate of reaction we may have an uncontrollable reactor. Fogler (1992) gives an example of how the dominant variables can change according to the rate controlling step in a slurry reactor. For example, when the gas-liquid mass transport controls the rate, the dominant variables are stirring rate and the partial pressure of reactant in the gas phase. On the other hand, when the chemical reaction governs the rate, the dominant variables are temperature, amount of catalyst, and liquid phase reactant concentration. We can only begin to imagine how well such a reactor would meet stated economic objectives if it operated in the gas mass transfer regime with a control system designed for kinetic control.

4.7.3 Design for partial control

Once the most appropriate reactor system has been chosen, the next important design issue to consider is how to provide enough manipulated variables to ensure adequate partial control. The first step in this

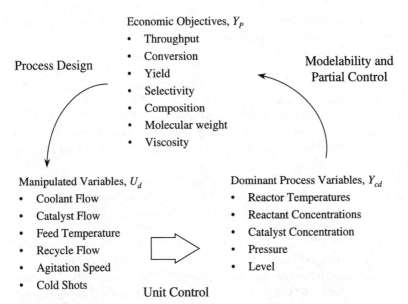

Economic Objectives, Y_p
- Throughput
- Conversion
- Yield
- Selectivity
- Composition
- Molecular weight
- Viscosity

Process Design

Modelability and Partial Control

Manipulated Variables, U_d
- Coolant Flow
- Catalyst Flow
- Feed Temperature
- Recycle Flow
- Agitation Speed
- Cold Shots

Unit Control

Dominant Process Variables, Y_{cd}
- Reactor Temperatures
- Reactant Concentrations
- Catalyst Concentration
- Pressure
- Level

Figure 4.34 Relationship among economic objectives, dominant variables, and manipulated variables.

procedure is to identify the dominant variables, since they are the ones that should be part of the primary control structure. It may not be necessary to control all dominant variables to stabilize the unit. For example, with sufficient hydrogen recycle, the HDA reactor can be operated by using only the inlet temperature for unit control. On the other hand, the autocatalytic plug-flow reactor described above requires more than inlet temperature control to hold the unit in a defined state. Once the minimum number of dominant variables is identified for stable unit control, we must check if this set is sufficient for partial control of the economic objectives. If not, we provide additional measured and manipulated variables to improve the performance. Figure 4.34 illustrates the relationship between economic objectives, dominant variables, and manipulated variables.

Arbel et al. (1997) give a detailed account of this procedure applied to a fluidized catalytic cracker (FCC). They show that unit control is possible when only one of the four dominant variables is under feedback control. The effectiveness of the partial control scheme is limited in satisfying the economic objectives when only one dominant variable is in closed-loop control. Superior reactor performance is achieved when all four dominant variables in the reactor are used. However, this requires manipulated variables that were not part of older FCC designs. The new manipulators have been added on modern units to make them more capable of meeting new economic objectives and added constraints. These new manipulators are the catalyst cooler in the regenerator and a feed heater to the riser reactor shown in Fig. 4.4. However, adding new manipulated variables requires both imagination and investment dollars. In some cases it may be obvious what to do, and it boils down to a trade-off between operating performance and investment cost. In other cases it may not even be clear how to add another manipulator regardless of what it might cost. It may be that a completely different reactor design must be considered to achieve the desired level of control.

Before we leave this section we would like to point out another interesting idea mentioned by Shinnar (1981). He suggested that when we are faced with model uncertainties at the design stage, it may be possible to *overdesign the control system*. What is meant by this is that we enlarge the set (Y_{cd}, U_s, U_d). For example, assume that we are designing a packed plug-flow reactor system where we are uncertain about the effects of backmixing or thermal conduction through the packing. When inlet temperature is the only controlled variable the possibility of multiple steady states and open-loop instability becomes a reality. To guard against these uncertainties we could consider alterations to the basic design that would allow us to control more dominant variables, in this case temperatures and compositions along the length of the reactor. We could use intermediate coolers or cold shots as the new manipulated variables required for control of the additional control points.

Similarly, the uncertainties around the rate controlling step for the slurry reactor may also be dealt with through overdesign of the control system. In that case it may be prudent to provide a couple of control loops that could control the addition of reactant gas through some high-pressure jet spargers or provide a variable speed agitator.

4.7.4 Design for responsiveness

When the unit control structure has been established, we would like to design the process such that the control loops are as responsive as possible. Interestingly enough, we can get clues on how to do this from the area of irreversible thermodynamics. The details are spelled out in Appendix A but let us give a brief introduction here, based on a very simple analog.

Assume we want to heat a pot of water on the stove from room temperature to a particular temperature below the boiling point. The time it takes to heat the water depends, in part, on how much heat Q_s is contained (stored) in the water when we reach the final temperature T_1:

$$Q_s = mC_P(T_1 - T_0) \qquad (4.30)$$

where m = mass of water in pot
C_P = heat capacity of water
T_0 = ambient temperature

The time for heating also depends on the supply of heat from the burner

$$\dot{Q}_s = UA\Delta T \tag{4.31}$$

where $\dot{Q}_s$ = rate of heat supply
U = heat transfer coefficient
A = heat transfer area
ΔT = temperature difference between burner and pot

Finally, the time for heating depends on rate of heat loss $\dot{Q}_d$ (dissipation) due to evaporation and convection. If we assume that the supply and dissipation rates remain constant (true only over narrow intervals of water temperature), the time t_h for heating the pot to the target temperature is

$$t_h = \frac{Q_s}{\dot{Q}_s - \dot{Q}_d} \tag{4.32}$$

As we all know, the heating time is reduced when there is little water in the pot, when the target temperature is low, when the burner is hot, and when there is a lid on the pot. Assume now that we took all those measures and managed to overshoot the target temperature. What do we do then? We turn off the burner, remove the lid, and probably stir to improve the rate of heat dissipation. The time it takes for the water to settle to the correct temperature is

$$t_c = \frac{Q_s^E}{\dot{Q}_d} \tag{4.33}$$

where t_c = cooling time and Q_s^E = excess heat stored in water. Once we arrived at the correct water temperature, we can hold it there by matching the supply to the dissipation rate, $\dot{Q}_s = \dot{Q}_d$.

It is intuitively clear in this example [and also seen in Eqs. (4.32) and (4.33)] that the response time is a direct function of the storage of heat and inversely proportional to the rate of heat removal or supply. From a control standpoint we can shorten the response time by affecting the supply and dissipation rates.

It turns out that the water pot example is a nice analog for a general process system such as a reactor. In the process system we characterize the energy storage with the thermodynamic state function *exergy B*, instead of heat Q_s. Heat is of course not a general thermodynamic state function but it plays the role of one in the water pot example. In the general process we replace the heat dissipation rate $\dot{Q}_d$, by the *exergy destruction rate* $T_0\dot{\sigma}$. The rate of exergy destruction is directly related to the rate of entropy production $\dot{\sigma}$ in the universe as a result of performing our process. We can show that the response rate of a process control loop is inversely proportional to the rate of entropy production caused by the manipulated variable. In other words, if we can find manipulated variables that strongly affect the rate of entropy production we can achieve responsive control.

We show in Appendix A that the rate of entropy production depends upon the product of fluxes and gradients. Fluxes and gradients are connected in the sense that a gradient is capable of generating a flux. For example, heat flows across temperature gradients and material flows across composition gradients. By manipulation of the fluxes or the gradients we can affect the rate of entropy production.

Temperature control of exothermic reactors involves manipulating the temperature gradient between the process and the coolant. Whenever we can affect the gradient significantly we are likely to have a responsive control loop. When the jacket temperature is affected by coolant flow, inlet temperature, or boiling point, it is easy to see that the flux can be changed more in designs with larger heat transfer areas. This follows directly from the heat transfer equation [e.g., Eq. (4.31)]. A small change in the coolant temperature will amount to a large relative change in ΔT and cause a major change in the heat flux. The product of heat flow and temperature gradient affects the entropy production rate and hence the time response for control.

Luyben and Luyben (1997) give several reactor examples where a large heat transfer area is beneficial for temperature control. The classic example is the scale-up of a jacketed CSTR. A small pilot plant reactor has a large heat transfer area compared to the reactor volume and temperature control is excellent. When the reactor is scaled to commercial size, the surface-to-volume ratio becomes unfavorable for control. The large heat release coupled with a relatively small UA forces the coolant rate to be high and the gradient ΔT to be large. Changes in the coolant rate now cause only minor relative changes to the gradient and the heat flux. The change in entropy production per change in cooling rate flow is small and control suffers.

Another example cited in Luyben and Luyben (1997) is when a large jacketed CSTR is replaced by several smaller CSTRs in series. For most reactions, a series of CSTRs has a lower total volume than a single CSTR for the same production rate and operating temperature. This smaller total reactor volume produces a smaller surface area and a larger ΔT, resulting in poor temperature control, particularly in the first reactor.

We have also seen the importance of a large heat transfer area in other examples given in this chapter. For example, we showed that the stability conditions for a cooled CSTR are

$$a_{11} + a_{22} > 0$$

$$a_{11}a_{22} - a_{12}a_{21} > 0$$

When we insert the proper terms from the Jacobian matrix we find the following set of inequalities must be satisfied for open-loop stability.

$$(V\rho C_P + M_w C_w) \cdot \frac{1 + \bar{k}\tau}{\tau} + F\rho C_P + UA >$$

$$\bar{k}\bar{C}_A V(-\Delta H)E_a/R\bar{T}^2 >$$

$$\frac{1 + \bar{k}\tau}{\bar{k}\tau} \cdot (-F\rho C_P - UA + \bar{k}\bar{C}_A V(-\Delta H)E_a/R\bar{T}^2)$$

We see by inspection that stability can always be ensured by making UA sufficiently large.

The principle of designing for small gradients is not limited to heat transfer examples. It applies to other thermodynamic gradients as well. For example, sparged reactors with fast reactions benefit from small gas bubbles with a large surface area to promote mass transfer. Under those circumstances minor variations in the partial pressure of reactants give a rapid response in overall reaction rates.

Chemical reactions are also influenced by gradients in the chemical potentials of the reactants and products. The chemical reaction gradient is called the *affinity*, A_j. The affinity for reaction j containing n components is

$$A_j = -\sum_{i=1}^{n} v_{ij}\mu_i$$

where v_{ij} = stoichiometric coefficient for component i in reaction j and μ_i = chemical potential of component i. The flux belonging to A_j is the rate of reaction, r_j.

Consider, for example, a simple reaction $A \to B$ with a rate expression $r_1 = kC_A$. The affinity for this reaction is

$$A_1 = -(-1 \cdot \mu_A + 1 \cdot \mu_B) = \mu_A^0 - \mu_B^0 + RT \ln \frac{C_A}{C_B}$$

A small affinity implies a low concentration of component A and a large concentration of component B, in other words, a high degree of conversion. If we consider controlling the reactor by adjusting the concentration of reactant we would get the fastest response in designs using large, backmixed reactors with nearly complete conversion of reactant. The worst response is obtained in small plug-flow systems with low per-pass conversion of reactant. A large one-pass reactor is easier to control than a smaller reactor in a recycle loop. We demonstrated this in Chap. 2. There we also showed that a reactor followed by a stripping column with recycle is cheaper than one large reactor. Again, this is an important example of the trade-off between steady-state investment cost and controllability.

4.8 Plantwide Control

So far we have dealt with control of reactors as isolated units. We now examine how reactors are controlled when they are part of an integrated plant. In principle, nothing new is introduced beyond the useful concepts involved in partial control. We can delineate among three cases:

1. All dominant variables are controlled at the unit level with manipulated variables local to the reactor.
2. Some dominant variables are controlled at the unit level.
3. The reactor is not controlled at the unit level.

Case 1 is desirable from the standpoint that it eliminates interactions from the rest of the plant. In other words, it is transparent to the reactor whether it is an isolated unit or part of a process with recycles. The economic objectives of the process are satisfied through partial control by adjusting the setpoints of the feedback loops. We can argue that the vinyl acetate reactor discussed in Chap. 11 falls in this category. The dominant variables are reactor exit temperature and oxygen inlet concentration. Both of these variables are controlled at the unit, making the reactor resilient against disturbances from the separation system.

Case 2 includes many of the example systems studied in this book. For example, reactors with temperature as the only controlled variable fall into this category. Also, the isothermal ternary scheme CS4 shown in Fig. 2.13a has a local composition controller on one of the dominant variables, the composition of component A. However, Case 2 is characterized by the fact that other dominant variables are not controlled at the reactor. Instead, the plantwide control structure plays a significant role in its ability to influence these uncontrolled variables. When the uncontrolled compositions become disturbances and the controlled dominant variables are too weak, we have difficulties. On the other hand, the plantwide control structure can be arranged to provide indirect control of the dominant composition variables, thereby augmenting the unit control loops. The HDA process provides a good illustration. The dominant variables are reactor inlet temperature and toluene composition. Only inlet temperature is controlled locally by the heat added to the furnace. If we let the toluene composition be a disturbance to the reactor we may have serious control difficulties especially with a small furnace. However, when we arrange the plantwide control loops for indirect control of the toluene feed concentration, we enlarge our opportunities for partial control. It is now possible to use a combination of setpoints for the inlet temperature and the toluene recycle to impact the economic objectives of throughput, yield, quality, etc.

Case 3, finally, provides the ultimate challenge for the plantwide control structure. Here, all the dominant variables in the reactor are influenced by the actions of controllers elsewhere in the plant. Now it becomes imperative that the plantwide controllers provide indirect control over all or most of the dominant variables. Several examples in Chap. 2 demonstrated this. As we showed in Chap. 2, it is very easy to configure schemes that turn the dominant variables into reactor disturbances. These schemes don't work at all. Consequently, we do not recommend building plants without local unit operation control for the reactor.

4.9 Polymerization Reactors

4.9.1 Basics

Polymers are long-chain molecules composed of repeated smaller units called *monomers*. The term *polymer* spans an enormous spectrum of substances that find widespread use in virtually all aspects of modern society. Polymers range from high-volume commodity types (polyethylene, polystyrene, etc.), to synthetic fibers (polyesters, polyamides, etc.), to engineering resins (polycarbonates, polyacetals, etc.), and beyond.

Polymerization reactors are generally one part of a large process involving monomer production and purification; polymer production; polymer recovery, isolation, and finishing; and monomer recovery and recycle. The general principles of plantwide control fit into continuous polymer processes because of this integration. Many of the concepts presented in this chapter directly apply to polymer reactors. As noted earlier, entropy decreases in polymerization reactions, which means that they are typically highly exothermic to satisfy Eq. (4.2). Reactor design for agitation and heat removal are crucial, particularly when the polymeric materials become highly viscous. In some cases the reactions are reversible at the normal operating temperatures, so the system ends up at equilibrium.

The basic concepts of modeling, open-loop behavior, reactor control, and plantwide control apply to polymer processes. It would be folly to attempt a comprehensive treatment of the subject in this text. We refer

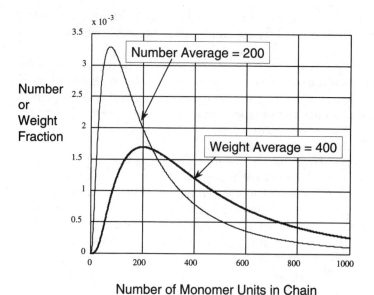

$$x\ 10^{-3}$$

Number or Weight Fraction

Number Average = 200

Weight Average = 400

Number of Monomer Units in Chain

Figure 4.35   Number and weight distributions for polymer.

the interested reader to Grulke (1994) and other excellent books on polymer science and engineering. However, we want to highlight in this section a key feature about polymerization reactors that is unique from other kinds of reactors.

The key feature of any polymer is that all chains are *not* of the same length. This gives rise to a *distribution* of various chain lengths (or molecular weights). This distribution can be characterized on the basis of number or weight fraction (Fig. 4.35). The number average (M_n) and weight average (M_w) molecular weights (or chain lengths) are the values where the areas under the distribution curves are equal to the left and right.

Certain important properties of polymers are directly related to the average chain length and the distribution. One of particular importance is the viscosity. This affects the flowability of the polymer and the kinds of applications where it can be used (injection molding, blow molding, fiber spinning, sheet formation, coating, etc.).

We are therefore concerned in polymer reactors to produce the desired average number of monomer units per chain and also the distribution (i.e., low and high molecular weight tails). The *polydispersity* PD is the measurement of the distribution and is the ratio of the weight average to number average molecular weights.

$$PD = \frac{M_w}{M_n} \tag{4.34}$$

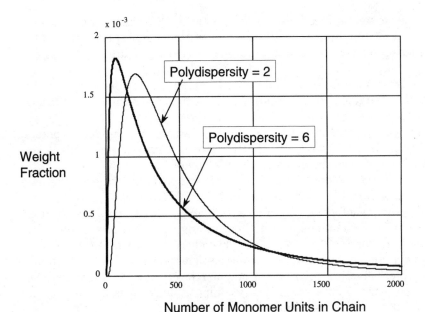

Figure 4.36   Spread of molecular weights.

A polymer may not have many high-molecular-weight molecules. They have little effect on the number average but contribute significantly to the weight average. As illustrated in Fig. 4.36, two polymers with the same weight average molecular weight can have completely different properties because of the difference in the distribution. The polymer with the lower number average molecular weight has a larger spread in the distribution, a longer high molecular weight tail, and a larger value of polydispersity. The *Flory distribution* is called the most probable and has a polydispersity equal to 2.

Other polymer properties also are important in addition to molecular weight and viscosity. If the polymer comes out of solution in the reactor as solid particles, then we would like to have the desired particle size distribution. Polymers composed of several different monomers, or *copolymers*, must have the appropriate compositions and segments along the polymer chain. The color of the polymer is important in some applications, as is the temperature of phase transitions, among other properties.

4.9.2   Dominant variables

Hence the key economic control objectives for polymer reactors are typically average molecular weight, polydispersity, viscosity, composition, partical size distribution, and production rate (plus color, phase transition temperature, etc.). It is often difficult (or impossible) to measure directly on line many of these important polymer properties. Many of these properties must be measured in a laboratory, introducing significant deadtime for these slow variables. This reinforces the need to use the ideas of dominant variables and partial control presented in this chapter. What are the dominant variables that affect these economic objectives?

Temperature is certainly a dominant variable for polymer reactors. Many of the reactor designs that have been discussed in previous sections can be used for polymerization. Any of the techniques discussed for heat removal apply here. However, temperature is never the only dominant variable in these systems.

Polymer reactors can often be a complex combination of many different physical phenomena (reaction, mixing, phase transfer, heat and mass transfer, etc.). Reactor design then becomes crucial to ensure that we have enough manipulators to achieve partial control of the dominant variables affecting the desired polymer properties. The new features for polymer reactors are typically composition, molecular weight, and molecular weight distribution.

For polymers that come out with the Flory distribution, we don't have a handle to control polydispersity. However, we almost always have a way to control conversion by using temperature, initiator or catalyst, or chain transfer agents. One equation that helps in looking at partial control is the relation between the degree of polymerization (DP) and conditions within the reactor:

$$DP = \frac{\text{rate of propagation}}{\text{rate of chain transfer} + \text{rate of termination}} \qquad (4.35)$$

DP is the average number of monomer (or repeat) units per polymer chain and so is directly related to molecular weight (or viscosity). This relationship shows that we must have control over variables that have a significant effect on propagation, chain transfer, and termination to achieve the desired polymer properties. What are these variables? They are the same as those we have discussed throughout this chapter: temperature, reactant monomer concentrations, concentrations of chain transfer agents or other impurities that affect polymerization, initiator or catalyst concentration, residence time, etc.

Polymer reactor control then boils down to controlling all dominant variables to setpoint using manipulators with a fast response and then adjusting the setpoints of the controlled variables to achieve the desired economic objectives. The trick is to determine the dominant variables and manipulators in addition to their relationships. Some key manipulators are heat removal (for externally cooled systems) or conversion control (for adiabatic systems) for temperature control; monomer, solvent, initiator, and catalyst feed stream flows (or compositions) to the reactor; and reactor effluent or feed flow to control level or residence time.

Plantwide issues of recycle and component inventory control play a significant role for polymer reactors. Because of their value, unconverted monomers are generally recovered from the polymer for recycle back to the reactor. These recycle streams most often contain impurities that can affect the polymerization (molecular weight, conversion, composition, color, etc.). In some cases a particular component impurity can be a dominant variable. If this impurity cannot be controlled and if there is no other equally dominant variable present that can be controlled, then the result will usually be an undesirable polymer product.

We are now going to discuss the two major types of polymerization systems, step growth and chain growth, and show what the difference implies for their control.

4.9.3 Step growth

Stepwise polymerization occurs from the intermolecular reaction of two different reactive end groups and the production of a low-molecular-weight "leaving group." Nylon 6,6 is an example of a step-growth polymer. The polyamide is made from the reaction of adipic acid and hexamethylene diamine, both of which have two reactive end groups. Water is the leaving group in this system.

$$HOOC - (CH_2)_4 - COOH + H_2N - (CH_2)_4 - NH_2$$

$$\rightleftharpoons HOOC - (CH_2)_4 - CONH - (CH_2)_4 - NH_2 + H_2O$$

This is an equilibrium reaction at typical operating conditions, which has several consequences for reactor design and control. Temperature control is of course important for its effect on the equilibrium conditions. In step-growth polymerization, several stages are often used to eliminate the volatile leaving group and allow the reaction to proceed to high conversion and molecular weight. For these systems material recycle is not typically a major factor.

For the liquid-phase reactor shown in Fig. 4.37, monomer feed is introduced and the effluent stream controls the level (residence time). Heat is removed via cooling water. We want to remove the water to push the equilibrium to the right and increase conversion. Due to its volatility, it would be natural to remove the water vapor from the reactor to control pressure.

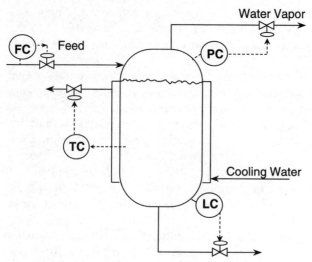

Water Vapor

FC Feed

PC

TC

Cooling Water

LC

Figure 4.37   Control of step-growth polymerization reactor.

4.9.4 Chain growth

Chain-growth polymerization does not produce a leaving group. Rather, it results from the coupling of reactive centers (often free radicals or ions) adding monomer units. Polyethylene is an example of a chain-growth polymer, where the propagation step is

$$R - CH_2 - CH_2^\bullet + CH_2 = CH_2 \rightarrow R - CH_2 - CH_2 - CH_2 - CH_2^\bullet$$

In this system conversion affects polymer properties. We typically cannot go to high conversion because of molecular weight or heat removal constraints (if adiabatic). There may also be a large increase in viscosity that affects the heat removal, agitation, and processability of the polymer solution. Here conditions dictate the kind of molecular weight distribution. The polymer is often affected by impurities and chain transfer agents that determine the amount of branching and termination.

For an adiabatic reactor, we may be able to control temperature and conversion using the initiator feed flow (Fig. 4.38). Incomplete conversion introduces recycle streams for the monomer. Because of the effect of chain transfer agents, we often must be able to measure the feed compositions to the reactor. Further, we must know what the chain transfer agents do if they are dominant variables to have any chance of controlling the molecular weight. So our control will only be as good as our correlations or models. Hence in polymer reactors we often have to use what is basically "steady-state" control on the setpoints of the dominant variables to achieve many of the control objectives that de-

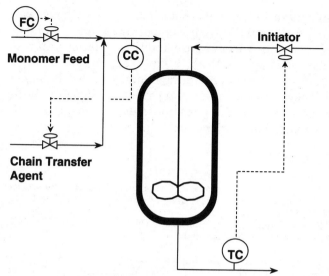

Figure 4.38 Control of adiabatic chain-growth polymerization reactor.

pend upon infrequent laboratory measurements (the notion of modelability in Fig. 4.34).

4.10 Conclusion

Here we have dealt with the control of chemical reactors. We covered
some of the fundamentals about kinetics, reactor types, reactor models,
and open-loop behavior. In particular we have shown that reactors with
recycle or backmixing can exhibit multiple steady states, some of which
are unstable. Nonlinearities in reactor systems also frequently give
rise to open-loop parametric sensitivity.

Most importantly, we introduced the ideas of dominance, effective
degrees of freedom, and partial control for chemical reactors. In essence,
dominant variables are controlled by manipulators with a fast response
and the setpoints are adjusted to achieve the economic objectives. These
notions are useful in this context, but they can be utilized more widely
for other unit operations.

Temperature is commonly the most dominant variable in reactor
systems. Since many chemical reactors are exothermic, controlling the
dominant variable in these systems amounts to removing the exothermic reaction heat through temperature control. We gave many examples of how that is done. In cases where temperature is not the most
dominant variable, compositions typically dominate. In this case unit
control is not localized to the reactor since composition control is affected by other parts of the plant.

Even though unit operation control is not addressed in our plantwide control design procedure until Step 8, it is important to understand up front what all the dominant variables are and their relationship with potential manipulators. This is particularly true if appropriate manipulators are unavailable, in which case design changes must be made. The dominant variables influence several steps in the design procedure, in particular our choice of controlling reactor temperature, production rate, and recycle stream compositions.

4.11 References

Arbel, A., Huang, Z., Rinard, I. H., and Shinnar, R. "Dynamics and Control of Fluidized Catalytic Crackers. 1. Modeling of the Current Generation of FCC's," *Ind. Eng. Chem. Res.*, **34**, 1228–1243 (1995a).

Arbel, A., Rinard, I. H., and Shinnar, R. "Dynamics and Control of Fluidized Catalytic Crackers. 2. Multiple Steady States and Instabilities," *Ind. Eng. Chem. Res.*, **34**, 3014–3026 (1995b).

Arbel, A., Rinard, I. H., and Shinnar, R. "Dynamics and Control of Fluidized Catalytic Crackers. 3. Designing the Control System: Choice of Manipulated and Measured Variables for Partial Control," *Ind. Eng. Chem. Res.*, **35**, 2215–2233 (1996).

Arbel, A., Rinard, I. H., and Shinnar, R. "Dynamics and Control of Fluidized Catalytic Crackers. 4. The Impact of Design on Partial Control," *Ind. Eng. Chem. Res.*, **36**, 747–759 (1997).

Bilous, O. and Amundson, N. R. "Chemical Reactor Stability and Sensitivity," *AIChE J.*, **1**, 513–521 (1955).

Bilous, O. and Amundson, N. R. "Chemical Reactor Stability and Sensitivity II. Effect of Parameters on Sensitivity of Empty Tubular Reactors," *AIChE J.*, **2**, 117–126 (1956).

Borio, D. O., Bucalá, V., Orejas, J. A., and Porras, J. A. "Cocurrently-Cooled Fixed-Bed Reactors: A Simple Approach to Optimal Cooling Design," *AIChE J.*, **35**, 1899–1902 (1989).

Bucalá, V., Borio, D. O., Romagnoli, J. A., and Porras, J. A. "Influence of Cooling Design on Fixed-Bed Reactors—Dynamics," *AIChE J.*, **38**, 1990–1994 (1992).

Denbigh, K. G., and Turner, J. C. R. *Chemical Reactor Theory*, 2d ed., New York: Cambridge University Press (1971).

Fogler, H. S. *Elements of Chemical Reaction Engineering*, 2d ed., Englewood Cliffs: Prentice-Hall (1992).

Froment, G. F., and Bischoff, K. B. *Chemical Reactor Analysis and Design*, New York: Wiley (1979).

Grulke, E. A. *Polymer Process Engineering*, Englewood Cliffs, N.J.: Prentice-Hall (1994).

Il'in, A., and Luss, D. "Wrong-Way Behavior of Packed-Bed Reactors: Influence of Reactant Adsorption on Support," *AIChE J.*, **38**, 1609–1617 (1992).

Il'in, A., and Luss, D. "Wrong-Way Behavior of Packed-Bed Reactors: Influence of an Undesired Consecutive Reaction," *Ind. Eng. Chem. Res.*, **32**, 247–252 (1993).

Luyben, W. L. *Process Modeling, Simulation and Control for Chemical Engineers*, 2d ed., New York: McGraw-Hill (1990).

Luyben, W. L., and Luyben, M. L. *Essentials of Process Control*, New York: McGraw-Hill (1997).

Metha, P. S., Sams, W. N., and Luss D. "Wrong-Way Behavior of Packed-Bed Reactors: 1. The Pseudo-Homogeneous Model," *AIChE J.*, **27**, 234–246 (1981).

Pinjala, V., Chen, Y. C., and Luss, D. "Wrong-Way Behavior of Packed-Bed Reactors: II. Impact of Thermal Dispersion," *AIChE J.*, **34**, 1663–1672 (1988).

Ray, W. H., and Hastings, S. P. "The Influence of the Lewis Number on the Dynamics of Chemically Reacting Systems," *Chem. Engng Sci.*, **35**, 589–595 (1980).

Shinnar, R. "Chemical Reactor Modeling—The Desirable and the Achievable," *ACS Symposium Series*, **72**, 1–36 (1978).

Shinnar, R. "Chemical Reactor Modeling for the Purposes of Controller Design," *Chem. Eng. Commun.*, **9**, 73–99 (1981).

Shinnar, R. Private communication (1997).

Shinnar, R., Doyle, F. J., Budman, H. M., and Morari, M. "Design Considerations for Tubular Reactors with Highly Exothermic Reactions," *AIChE J.*, **38**, 1729–1743 (1992).

van Heerden, C. "Autothermic Processes—Properties and Reactor Design," *Ind. Eng. Chem.*, **45**, 1242–1247 (1953).

van Heerden, C. "The Character of the Stationary State of Exothermic Processes," *Chem. Engng Sci.*, **8**, 133–145 (1958).

Vleeschhouwer, P. H. M., Garton, R. D., and Fortuin, J. M. H. "Analysis of Limit Cycles in an Industrial Oxo Reactor," *Chem. Engng Sci.*, **47**, 2547–2552 (1992).

5

Heat Exchangers
and Energy Management

5.1 Introduction

Step 3 of our plantwide control design procedure involves two activities. The first is to design the control loops for the removal of heat from exothermic chemical reactors. We dealt with this problem in Chap. 4, where we showed various methods to remove heat from exothermic reactors and how to control the temperature in such reactors. At that point we assumed that the heat was removed directly and permanently from the process (e.g., by cooling water). However, it is wasteful to discard the reactor heat to plant utilities when we need to add heat in other unit operations within the process. Instead, a more efficient alternative is to heat-integrate various parts of the plant by the use of process-to-process heat exchangers.

The second activity in Step 3 therefore deals with the management of energy in heat-integrated processes. The reason we look at energy management in conjunction with heat removal controls for exothermic reactors is that these issues are often closely coupled. For example, it is common to preheat the feed to adiabatic tubular reactors with the reactor effluent stream. This recovers heat from the reactor that otherwise would be discarded to utilities. However, it also retains the heat within the process and could cause excessive temperatures around the reactor unless properly managed. Another potential problem is the thermal feedback provided by the feed-effluent exchanger. As we saw in Chap. 4, such feedback can cause multiplicity and open-loop instability of the reactor. The reactor control system must therefore be revisited in light of the heat integration schemes considered.

Heat management is accomplished by controlling the flow of energy in various heat exchangers. We differentiate between utility exchangers and process-to-process heat exchangers. Utility exchangers allow us to import heat from heat sources or discard heat to the environment. Process-to-process exchangers, on the other hand, are used to transfer heat from one part of the process to another. We will look at the unit operation controls of both types of exchangers.

We will not discuss the steady-state design techniques involved in arriving at a heat-integrated process. This is adequately covered in texts on process design (e.g., Douglas, 1988). We will, however, examine the dynamic and control implications of such integrated designs. In particular, we will point out design modifications required to make a heat-integrated plant operable and controllable.

5.2   Fundamentals of Heat Exchangers

Process understanding precedes control system design. The area of heat management is no exception. To provide the proper background for this chapter we therefore start with a brief review of heat exchanger characteristics and the thermodynamic foundations of heat management. We examine the thermal efficiency of a typical chemical process and explore steady-state incentives for heat conservation and integration. We also extend the discussion from Chap. 4 regarding the role of thermodynamics in process control.

5.2.1   Steady-state characteristics

Figure 5.1 shows a typical shell-and-tube exchanger used to exchange sensible heat between two streams. The familiar design equations are

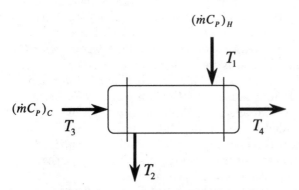

Figure 5.1   Schematic of a shell-and-tube heat exchanger.

$$Q = (\dot{m}C_P)_H(T_1 - T_2) = (\dot{m}C_P)_C(T_4 - T_3) \qquad (5.1)$$

$$= UA\frac{(T_1 - T_4) - (T_2 - T_3)}{\ln\dfrac{T_1 - T_4}{T_2 - T_3}} \qquad (5.2)$$

where
$$\begin{aligned}
Q &= \text{heat transferred} \\
T_1, T_2, T_3, T_4 &= \text{stream temperatures} \\
(\dot{m}C_P)_H &= \text{product of flowrate and specific heat capacity for} \\
&\quad \text{hot stream} \\
(\dot{m}C_P)_C &= \text{product of flowrate and specific heat capacity for} \\
&\quad \text{for cold stream}
\end{aligned}$$

Equations (5.1) and (5.2) are ideally suited to design the exchanger and thereby calculate the heat transfer area required to meet the specified stream temperatures. However, the equations are less convenient as rating equations to determine the steady-state exit temperatures for a given exchanger under different operating conditions. This is of course what we need in control studies when we must determine the process gain and the steady-state effects of disturbances. For those applications it is easier to use the effectiveness method (Gebhart, 1971; Jones and Wilson, 1997). The effectiveness of an exchanger is defined as

$$\epsilon = \frac{(\dot{m}C_P)_H(T_1 - T_2)}{(\dot{m}C_P)_{\min}(T_1 - T_3)} = \frac{(\dot{m}C_P)_C(T_4 - T_3)}{(\dot{m}C_P)_{\min}(T_1 - T_3)} \qquad (5.3)$$

where $(\dot{m}C_P)_{\min}$ is the smaller of $(\dot{m}C_P)_H$ and $(\dot{m}C_P)_C$.

For given values of the inlet flows and temperatures, the exit temperatures are explicitly calculated for a known exchanger effectiveness:

$$T_2 = T_1 - \frac{\epsilon(\dot{m}C_P)_{\min}(T_1 - T_3)}{(\dot{m}C_P)_H} \qquad (5.4)$$

$$T_4 = T_3 + \frac{\epsilon(\dot{m}C_P)_{\min}(T_1 - T_3)}{(\dot{m}C_P)_C} \qquad (5.5)$$

It can be shown that the effectiveness is determined by the exchanger's design parameters through the following equation:

$$\epsilon = \frac{1 - e^{-(1-r)N_{TU}}}{1 - re^{-(1-r)N_{TU}}} \qquad (5.6)$$

where $r = \dfrac{(\dot{m}C_P)_{\min}}{(\dot{m}C_P)_{\max}}$

$N_{\text{TU}} = \dfrac{UA}{(\dot{m}C_P)_{\min}}$

When the two streams exchanging heat have the same mass flowrate and heat capacity, $r = 1$, and the effectiveness is calculated from

$$\epsilon = \frac{UA/\dot{m}C_P}{1 + UA/\dot{m}C_P} \tag{5.7}$$

5.2.2 Heat exchanger dynamics

Heat exchangers have fast dynamics compared to other unit operations in a process. Normally the time constant is measured in seconds but could be up to a few minutes for large exchangers. Process-to-process exchangers should be modeled rigorously by partial differential equations since they are distributed systems. This introduces the correct amount of deadtime and time constant in the exit stream temperatures, but the models are inconvenient to solve. We have found that for the purpose of plantwide control studies it is not necessary to have such detailed descriptions of the exchanger dynamics, since these units rarely dominate the process response. Instead, it is often possible to construct useful models by letting two sets of well-stirred tanks in series exchange heat. This simplifies the solution procedure. Alternatively, we can use the effectiveness method to calculate the steady-state exchanger exit temperatures and then delay these temperatures by first-order time constants to capture the dynamics.

5.3 Thermodynamic Foundations

5.3.1 Energy, work, and heat

There are two aspects of energy that we shall be interested in, heat and work. Work is the hard currency of energy and it can always be completely converted into heat. Heat, on the other hand, is a less organized source of energy and it can only partly be converted into work. The remaining heat has to be discarded into a low-temperature heat sink. The largest fraction of work that can be extracted from a heat unit depends upon the absolute temperatures of the heat source and the heat sink.

$$\left(\frac{w}{q}\right)_{max} = \frac{T_1 - T_2}{T_1} \tag{5.8}$$

where w = work
          q = heat
          T_1 = absolute temperature of heat source
          T_2 = absolute temperature of heat sink

Since the work portion of energy is independent of the method of delivering the energy, it is most meaningful to compare the energy usage of various unit operations on a work-equivalent basis. We can then differentiate between work producers and work consumers. As a matter of generalization, we can state that it is possible, in principle, to extract work from spontaneous processes and that work is always needed to reverse such processes. When we cannot extract work from a spontaneous process, the work potential is lost.

What, then, are the spontaneous processes in a chemical plant? Here is a short but complete list.

- Chemical reactions
- Mixing of streams with different compositions, temperatures, and pressures
- Flow of material from high to low pressure
- Flow of heat from high to low temperature

We are rarely able to extract much work from chemical reactors. We mostly take out energy from the reactor in the form of heat. Also, when we let a hot stream heat up a cold stream in a heat exchanger we lose some of the work potential of the hot stream. The same is true for mixing and material flow across a pressure drop; we take advantage of the spontaneity of the process and make no attempt to recover work from it. In contrast, we often want to perform operations that are the reverse to the spontaneous direction. This always requires work. For example, separation is the opposite of mixing. The work demand of separating an ideal mixture of n components into pure products at constant temperature T is

$$w = -RT \sum_{i=1}^{n} x_i \ln x_i \qquad (5.9)$$

where w = separation work per mole of mixture
R = universal gas-law constant
T = absolute temperature of the separation
x_i = mole fraction of component i in the mixture

When we consider that we lose work potential in chemical reactors, in heat exchangers, and in mixing operations and combine this with the need for work in separation processes, pumping, and compression, it becomes clear that chemical processes are usually very inefficient from an energy standpoint. Of course, energy conservation is usually of secondary importance in chemical processing; safety, quality, and productivity are more important. Nevertheless, it is economically sound to try to improve the thermal efficiency of the plant. We just have to do it in a way that does not jeopardize the primary goals of the operation.

5.3.2 HDA example

To appreciate better the use of energy in a chemical plant and the opportunities that exist for heat integration, we will look at a specific case. The HDA process introduced in Chap. 1 will serve as our example. We recall the main reaction in this process is

$$\text{Toluene} + H_2 \rightarrow \text{benzene} + CH_4$$

Standard state heat of reaction $\Delta H^0 = -42.2$ kJ/mol

Change in Gibbs free energy $\Delta G^0 = -43.1$ kJ/mol

As mentioned in Chap. 4, ΔH^0 is the amount of heat that has to be removed to operate a reactor isothermally at the standard states of the components. The Gibbs free energy of reaction, ΔG^0, on the other hand, is how much work we could extract in the form of electrical energy from a hypothetical reactor operating as a reversible, isothermal fuel cell. Since we don't know how to build fuel cells around arbitrary reactions we usually extract the reaction energy as heat. Some of this heat can, of course, be converted to mechanical or electrical energy by use of a turbine and a generator. However, we take a healthy cut in our dividends. For example, let's say we made 500 psia steam from the hot effluents of the HDA reactor. Equation (5.8) gives us the maximum turbine work we could extract from the steam.

$$w \approx 42.2 \cdot \frac{515 - 298}{515} = 17.8 \text{ kJ/mol toluene}$$

Notice that we have used the standard state heat of reaction to estimate the heat evolved at 515 K. This is only approximately true since the standard state heat of reaction should be adjusted for the difference in heat capacities between the products and the reactants at the high temperature. The only point we are trying to convey is that the available work is significantly reduced when we use steam to drive a turbine. Furthermore, because of inefficiencies of the turbine and the generator, the electrical energy generated this way ends up as a small fraction of the 43.1 kJ/mol toluene that is theoretically available from the reaction.

We next turn to the separation section of the process. Equation (5.9) can help us get a rough estimate of what the separation costs would be. We first look at the total flow exiting the reactor and consider how much work would be required to separate this stream into pure components. Next, we consider the work associated with the vapor recycle stream. This work is subtracted from the total work of the exit stream since we don't have to separate the components in the gas recycle. When we use the flowrate and composition data from Chap. 10 and we assume that the separation is being done at a constant temperature (298 K), we get the following results by using Eq. (5.9):

$$w_{\text{Rx Exit}} = -F \cdot R \cdot T \cdot \sum_{i=1}^{5} x_i \ln x_i$$

$$= -553 \text{ mol/s} \cdot 8.31 \text{ J/mol K} \cdot 298 \text{ K}$$

$$\cdot \, [0.364 \ln (0.364) + 0.546 \ln (0.546) + 0.069 \ln (0.069)$$

$$+ \, 0.019 \ln (0.019) + 0.002 \ln (0.002)]$$

$$= 553 \cdot 8.31 \cdot 298 \cdot 0.97 = 1.3 \times 10^6 \text{ J/s}$$

$$w_{\text{Recycle}} = 444 \cdot 8.31 \cdot 298 \cdot 0.71 = 0.78 \times 10^6 \text{ J/s}$$

$$w_{\text{Separation}} = (1.3 - 0.78) \times 10^6 = 0.52 \text{ MW}$$

The net work of 0.52 MW for the separation section assumes that we perform all separations mechanically, i.e., with compressors and semipermeable membranes. In reality we use evaporation, partial condensation, and distillation to separate the components. We can use Eq. (5.8) to estimate how much heat, from 75 psia steam, we must supply to an ideal separation device to provide the necessary separation work. Again, this is only a rough estimate of the required heat.

$$q \approx w \cdot \frac{T_1}{T_1 - T_0} = 0.52 \cdot \frac{426}{426 - 298} = 1.7 \text{ MW}$$

In reality, partial condensers and distillation columns are far from ideal separation devices. There is plenty of mixing and heat degradation in these operations to lower the efficiency significantly. In fact, we can see from Table 10.6 in Chap. 10 that the net heat consumption for partial condensation and distillation is over 5 MW.

We can summarize the HDA example as follows. The process converts 132 kmol/h of toluene. If it were possible to extract all the work contained in the reactants we could generate 1.58 MW of electric power at standard conditions. When forced to generate power from steam we would obtain much less than 0.65 MW of electric power. When none of the reaction energy is converted to work we need to dissipate about 1.55 MW of heat to utilities. On the separation side we need 0.52 MW of mechanical power at standard conditions. If ideal heat-driven separation devices were available they would consume roughly 1.7 MW of heat. In actual distillation columns the separation cost is over 5 MW of heat.

This example shows clearly why most chemical processes are so thermally inefficient. In theory we have an excess of work available $(1.58 - 0.52 > + 1$ MW$)$. In reality we have a deficiency in heat $(1.55 - 5 < - 3.45$ MW$)$.

So far we have looked at the energy requirements of the process operating at ambient conditions (i.e., 25°C, 1 atm). However, these conditions are impractical other than for raw material and product storage. For example, the rate of reaction between toluene and hydrogen is not appreciable until the reactor inlet conditions are around 600°C and over 30 atm. However, this high temperature is impractical for material storage and for distillation. Distillation columns are mostly operated in the 50 to 200°C range since steam is commonly used in the reboilers and cooling water in the condensers. The bottom line is that, for practical reasons, we need to operate various processing steps at different conditions. The transition of process streams between these different conditions costs energy. To visualize just how much we need for this purpose we refer to the temperature-enthalpy diagram (T-H diagram) for the HDA process (Terrill and Douglas, 1987a; Douglas, 1988). A T-H diagram shows how much heat is required to bring the cold streams in the process to higher temperatures and how much cooling is needed to bring the hot streams down to lower temperatures. We find that as much as 26 MW of heating and cooling are required to transition streams within the HDA plant!

5.3.3 Heat pathways

It is important to realize that there are no thermodynamic restrictions on the energy requirement to transition streams between unit operations. In other words, the heating and cooling of streams is done for practical reasons and not to satisfy the laws of thermodynamics. This energy would not be an issue if all the processing steps operated at the same constant temperature. Furthermore, since raw materials and products are stored at roughly the same temperature, the net energy requirement for heating and cooling equals the heat losses from the process. We therefore realize that most of the energy required for heating certain streams within the process is matched by a similar amount required for cooling other streams. Heat recovered from cooling a stream could be recycled back into the process and used to heat another stream. This is the purpose of heat integration and heat exchanger networks (HENs).

From a plantwide perspective we can now discern three different

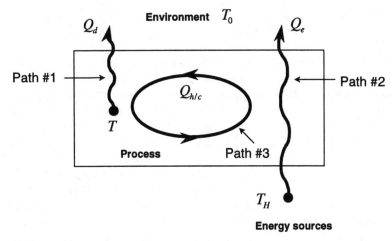

Figure 5.2 Heat pathways.

"heat pathways" in the process. See Fig. 5.2 for an illustration. The first pathway dissipates to the environment heat generated by exothermic reactions and by degradation of mechanical work (e.g., compression, pressure drop, and friction). This pathway is from inside the process and flows out. It is of course possible to convert some of the heat to work as it is removed from high temperatures in the process.

A second pathway carries heat from utilities into the process. Mechanical work is extracted from the heat as it flows from a high supply temperature to the lower temperature of the environment. This pathway goes through the process and is needed to satisfy the thermodynamic work requirements of separation. Work is also extracted from the heat stream to overcome process inefficiencies associated with stream mixing and heat transfer.

The third pathway is internal to the process. Here heat flows back and forth between different unit operations. The magnitude of this energy path depends upon the heating and cooling needs and the amount of heat integration implemented. Whenever the internal path is missing, and there is a heating requirement, the heat has to be supplied from utilities. The same amount of heat must eventually be rejected to the environment elsewhere in the process.

5.3.4 Heat recovery

We have tried to illustrate why chemical processes are thermally inefficient. First, the chemical work available in the reactants is dissipated as heat. Second, the work required for separation is usually supplied as heat to distillation columns, which have internal inefficiencies. Finally, energy is needed for heating and cooling functions that are independent of thermodynamic constraints. This all adds up to a low thermal efficiency.

Fortunately, we can make great improvements in the plant's thermal efficiency by recycling much of the energy needed for heating and cooling process streams. It is also possible to introduce heat integration schemes for distillation columns to reduce the separation heat. And finally we can recover the reaction heat in waste heat boilers and use the steam for power generation. There is of course a capital expense associated with improved efficiency but it can usually be justified when the energy savings are accounted for during the lifetime of the project. Of more interest to us in the current context is how heat integration affects the dynamics and control of a plant and how we can manage energy in plants with a high degree of heat recovery. We hope to address these concerns adequately in the remainder of this chapter. But before we embark on heat exchanger and energy management control methods we extend the discussion from Chap. 4 regarding thermodynamics and process control.

5.3.5 Exergy destruction principle

In Chap. 4 we introduced the concept of exergy in connection with the design of responsive control loops. Exergy is the mechanical work potential of a stream or the energy stored in a process relative to the conditions in the earth's environment. It incorporates all the work functions we have looked at so far, that is, work from heat [Eq. (5.8)], separation work [Eq. (5.9)], and the Gibbs free energy change of reaction. As opposed to plain "work," exergy is a thermodynamic state function (see Appendix A for more details on this). The reason exergy is a state function, and work is not, comes from the fact that exergy expresses the work potential of a stream *relative* to the equilibrium state of the environment. Without such a reference, the work portion of energy is undetermined.

Exergy and energy are expressed in the same units but the two entities are different in one important aspect. Energy is a conserved property, exergy is not. In fact, all spontaneous processes are accompanied by a degree of exergy destruction. While this property is important in identifying the thermal efficiencies of a process, it also has an interesting implication for process control. We showed in Chap. 4, and discuss further in Appendix A, that responsive control is achieved when the control system can significantly *alter the rate of exergy destruction* in the process. A limited amount of exergy is destroyed when energy is kept within the process by means of heat transfer between hot and cold streams. A maximum amount of exergy is destroyed when heat is dissipated from the process to the environment.

We can therefore make the general prediction that heat-integrated plants whose control systems manage energy by transferring excess heat from one location to another are not likely to be very responsive or controllable. On the other hand, when we provide means for the control system to discard excess heat to the environment through utility exchangers we can significantly improve the controllability of the heat-integrated plant. These auxiliary utility exchangers are needed only for control and will not show up on the original steady-state flowsheet. It is the responsibility of the control engineer to ensure that the auxiliary exchangers are incorporated in the design. They are another important example of the trade-off between steady-state economics and controllability.

While the exergy destruction principle has its roots in irreversible thermodynamics, we can understand its implications from intuitive grounds. Heat integration provides the same type of coupling between units that occurs with material recycle. A disturbance in one unit is recycled to another unit which in turn can affect the first unit. In material recycle loops we limit the effects of flow disturbances by flow controlling somewhere in the loop. Composition disturbances are harder to deflect and here we mostly rely on the unit operation controls and the plantwide control strategy to minimize their propagation. Energy disturbances will also be recycled unless they can be deflected by the control system. Fortunately there is a simple way of doing this. We can sense the disturbance, typically as a change in temperature somewhere in the loop, and divert the disturbance to the utility system by way of utility exchangers.

5.4   Control of Utility Exchangers

Armed with the thermodynamic fundamentals of heat management, we now take a closer look at the unit operation control loops for heat exchangers. We start with utility exchangers. These are used when heat is supplied to, or removed from, the process. Examples are steam-heated reboilers, electric heaters, fuel-fired furnaces, water-cooled condensers, and refrigerated coolers.

The purpose of unit operation controls is to regulate the amount of energy supplied or removed. This is typically done by measuring a temperature in the process and manipulating the flowrate of the utility. A PI-controller is adequate in most instances although derivative action can be used to compensate for the lag introduced by the thermowell. The location of the temperature measurement depends upon the purpose of the heat exchange. For example, when reaction heat is dissipated through a water cooler, the controlled temperature is in the reactor. We gave many examples of this in Chap. 4. Similarly, when the utility exchanger delivers heat to a separation system, the control point should be

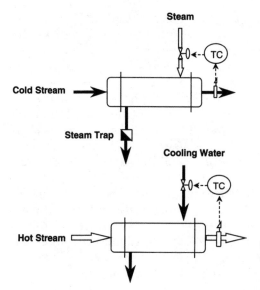

Figure 5.3 Control of utility exchangers.

located where the effects of the added energy are felt the most. Control of a tray temperature in a distillation column by manipulation of reboiler steam is a good example. Finally, when the utility exchanger is used for stream heating and cooling, the control point is on the stream being heated or cooled. Figure 5.3 shows two examples of the latter application.

In some situations the heat load on a utility exchanger is controlled without adjusting the utility flow. Consider the partial condenser on a distillation column shown in Fig. 5.4. The cooling water valve is run

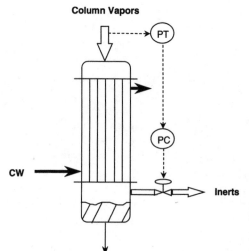

Figure 5.4 Indirect control of heat transfer rate in partial condenser with inerts.

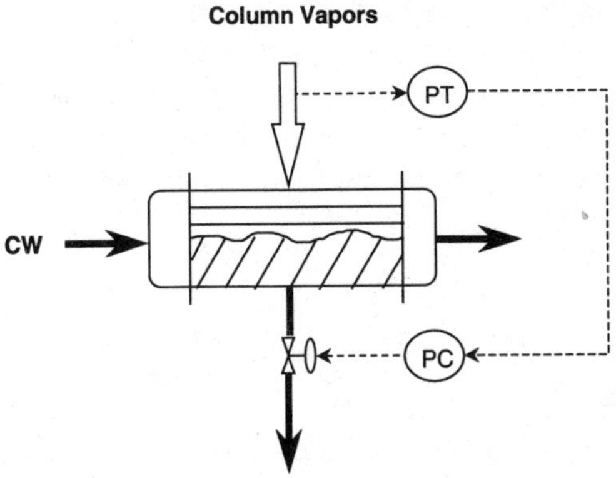

Column Vapors

Figure 5.5 Heat transfer control by variable heat transfer area in flooded condenser.

wide open and heat transfer is affected by adjusting the inert concentration in the condenser. This arrangement ensures the best recovery of condensables in the presence of inerts. When there are no inerts present in the column, heat transfer control can be done by flooding the condenser as in Fig. 5.5. By manipulating the condensate flow, the heat transfer area is adjusted to provide the desired amount of heat transfer for condensation at a constant cooling water flow. However, the final temperature of the condensate (the degree of subcooling) will vary with the condensing load.

Occasionally utility exchangers need no control loops at all. For example, the distillation column in Fig. 5.6 uses an uncontrolled cooling water condenser. The water-cooled condenser reduces the expensive

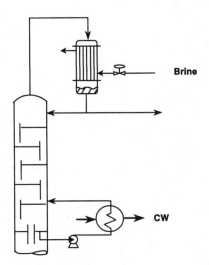

Brine

CW

Figure 5.6 Utility exchanger without heat transfer control.

duty of the top refrigerated condenser. However, the water-cooled exchanger cannot condense all the vapors. We minimize the refrigeration load by keeping the cooling water valve fully open and exchange as much heat as we can at the higher temperature. The remaining heat has to be taken out by the refrigerated condenser which, of course, has to be controlled somehow.

5.5 Control of Process-to-Process Exchangers

Process-to-process (P/P) exchangers are used for heat recovery within a process. Equations (5.4) and (5.5) tell us that we can control the two exit temperatures provided we can independently manipulate the two inlet flowrates. However, these flowrates are normally unavailable for us to manipulate and we therefore give up two degrees of freedom for temperature control compared to the case where we use utility exchangers on both streams. We can restore one of these degrees of freedom fairly easily. For example, it is possible to oversize the P/P exchanger and provide a controlled bypass around it as in Fig. 5.7a. It is also possible to combine the P/P exchanger with a utility exchanger as in Fig. 5.7b.

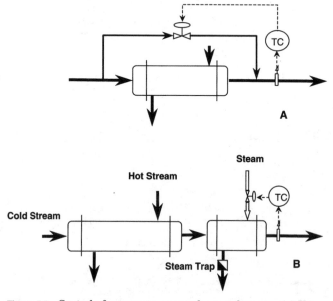

Figure 5.7 Control of process-to-process heat exchangers. (a) Use of bypass; (b) use of auxiliary utility exchanger.

5.5.1 Bypass control

When the bypass method is used for unit operation control, we have several choices about the bypass location and the control point. Figure 5.8 shows the most common alternatives. We may ask "Which option is the best?" It depends on how we define "best." As with many other examples in this book, it boils down to a trade-off between design and control. Design considerations might suggest we measure and bypass on the cold side since it is typically less expensive to install a measurement device and a control valve for cold service than it is for high-temperature service. Cost considerations would also suggest a small bypass flow to minimize the exchanger and control valve sizes. From a control standpoint we should measure the most important stream, regardless of temperature, and bypass on the same side as we control (e.g., Fig. 5.8a and c). This minimizes the effects of exchanger dynamics in the loop. We should also want to bypass a large fraction of the controlled stream since it improves the control range. This requires a large heat exchanger.

There are several general heuristic guidelines for heat exchanger bypass systems. However, this very much remains an open research area since these guidelines are not always adequate to deal with all of

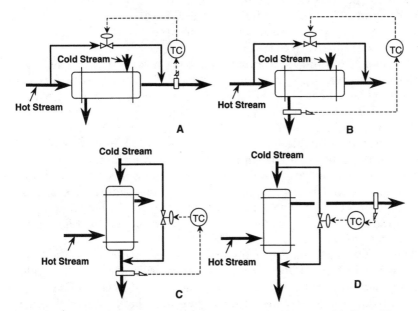

Figure 5.8 Bypass control of process-to-process heat exchangers. (a) Controlling and bypassing hot stream; (b) controlling cold stream and bypassing hot stream; (c) controlling and bypassing cold stream; (d) controlling hot stream and bypassing hot stream.

the issues for bypass systems. We typically want to bypass the flow of the stream whose temperature we want to control. The bypass should be about 5 to 10 percent of the flow to be able to handle disturbances. Finally, we must carefully consider the fluid mechanics of the bypass design for the pressure drops through the control valves and heat exchanger. Instead of using just one valve in the bypass line, we also can consider two valves or a three-way valve to provide sufficient rangeability of flows (Jones and Wilson, 1997).

5.5.2 Use of auxiliary utility exchangers

When the P/P exchanger is combined with a utility exchanger, we also have a few design decisions to make. We must first establish the relative sizes between the recovery and the utility exchangers. From a design standpoint we would like to make the recovery exchanger large and the utility exchanger small. This gives us the most heat recovery, and it is also the least expensive alternative from an investment standpoint. However, a narrow control range and the inability to reject disturbances make this choice the least desirable from a control standpoint.

Next, we must decide how to combine the utility exchanger with the P/P exchanger. This could be done either in a series or parallel arrangement. Physical implementation issues may dictate this choice but it could affect controllability. Finally, we have to decide how to control the utility exchanger for best overall control performance. An example will illustrate these issues.

Consider a distillation column that uses a large amount of high-pressure steam in its thermosiphon reboiler. To reduce operating costs we would like to heat-integrate this column with the reactor. A practical way of doing this is to generate steam in a waste heat boiler connected to the reactor as suggested by Handogo and Luyben (1987). We can then use some or all of this steam to help reboil the column by condensing the steam in the tubes of a stab-in reboiler. However, the total heat from the reactor may not be enough to reboil the column, so the remaining heat must come from the thermosiphon reboiler that now serves as an auxiliary reboiler. The column tray temperature controller would manipulate the steam to the thermosiphon reboiler. See Fig. 5.9 for the heat recovery arrangement.

The control performance of the heat-integrated reactor/column system shown in Fig. 5.9 deteriorates as the auxiliary reboiler provides less and less heat to the column. The reason is that uncontrolled variations in the steam pressure of the waste heat boiler affect the heat supplied to the column. When these variations are of the same order of magnitude as the total heat supplied by the auxiliary reboiler, the latter cannot compensate properly for the variations. Part of the prob-

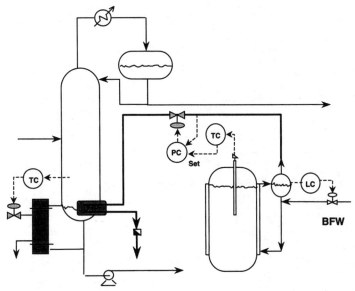

Figure 5.9 Reactor/column heat integration with auxiliary reboiler in parallel.

lem in this parallel arrangement is that the disturbances propagate into the column before the auxiliary reboiler has a chance to react. We can improve on this situation by providing a total heat controller. The principle of total heat control is simple. We measure the combined heat input from all heat sources. This becomes the input to the so-called Q controller that manipulates the utility valve. The total heat demand is adjusted by changing the Q controller's setpoint. See Fig. 5.10.

As a final note on the reactor/column example, we might have consid-

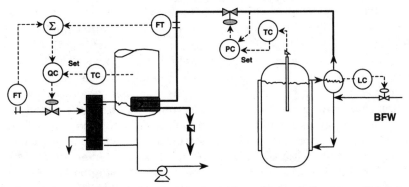

Figure 5.10 Reactor/column heat integration with auxiliary reboiler in parallel and Q controller.

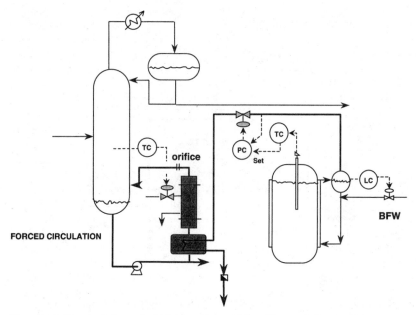

Figure 5.11 Reactor/column heat integration with auxiliary reboiler in series.

ered a series arrangement as in Fig. 5.11. Temperature variations from the reactor would now be attenuated in the auxiliary heater by virtue of changes in the temperature driving force for heat transfer.

5.6   Plantwide Energy Management

5.6.1   Introduction

We now look at the plantwide control issues around energy management. We need to identify the various pathways for heat and to devise a control strategy that allows effective delivery and removal of energy where it is needed with minimal propagation of disturbances. The HDA process continues to serve as our example. We previously noted that a practical implementation of this process requires far more energy in heating and cooling internal streams than is either generated in the reactor or used to drive the distillation columns. Therefore, when we look at efficiency opportunities in the HDA process it is fairly clear that we need to start with the heating and cooling requirements before considering other options. Indeed, the most obvious heat integration scheme is the one presented as the base case by Douglas (1988) and shown in Chap. 1 in this book. Here we use the hot effluent gases from the reactor to help preheat the recycle streams and the fresh feeds before they enter the reactor.

This simple design cuts down the energy consumption to one-third of what it would be in the worst case if all process streams had to be heated and cooled with utilities. We may even ask "Is it worth going beyond this point?" Terrill and Douglas (1987b) answered this question by studying six different energy-saving alternatives to the base case. The simplest of these designs (Alternative 1) recovers an additional 29 percent of the base case heat consumption by making the reactor preheater larger and the furnace smaller. The most complicated of the designs (Alternative 6) recovers 43 percent of the base case's net energy consumption. We showed a schematic of this highly integrated alternative in Chap. 2. Below we will outline how such a complex scheme might be controlled. Not too surprisingly we will find that it requires several design modifications to the process to ensure controllability and operability. These control-related changes may easily offset the steady-state savings from the high level of heat integration. Therefore, as a more viable alternative to the base case we will take a closer look at Alternative 1. This scheme represents a commonly used approach to heat integration. Its main advantage is that it confines the recycle of heat to the reactor. The disadvantage is that it provides a high degree of thermal feedback to the reactor which can be detrimental to the reactor's behavior (e.g., output multiplicity and open-loop instability).

5.6.2 Controlling plantwide heat integration schemes

In this section we outline the approach we would take in controlling a complex heat-integrated scheme like Alternative 6 of the HDA process. The first step is to identify the three heat pathways.

Path 1 is intended to carry heat from exothermic reactions that must be dissipated from the process. This path, shown in Fig. 5.12, goes through all three distillation column reboilers as well as the three preheaters before it terminates in the water cooler. An enthalpy disturbance in the reactor propagates through the entire plant.

Path 2, shown in Fig. 5.13, conveys heat to the distillation columns. This heat covers the thermodynamic work requirement for the separations. In Alternative 6 for the HDA process all the heat needed to run the process is supplied by the furnace. Heat intended for the columns must travel through the reactor and the preheaters before it reaches its destinations in the three condensers. The columns can indirectly upset the reactor.

Path 3 is the heating and cooling circuit that starts from the reactor exit and goes through the preheaters and column reboilers. In the preheaters the hot streams give up most of their enthalpy to the incoming cold feed streams that travel back to the reactor. This path is shown

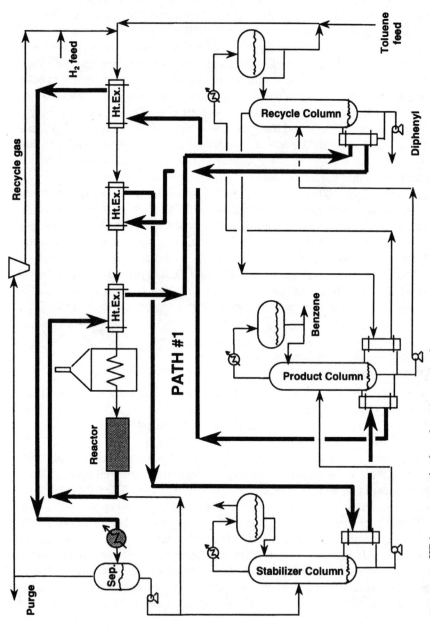

Figure 5.12 HDA process path of exothermic reaction heat.

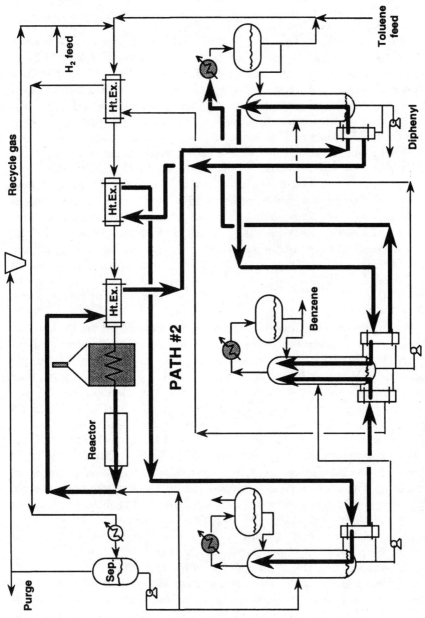

Figure 5.13   HDA process path of heat supply to distillation columns.

in Fig. 5.14. Like the other two paths in this alternative, the third path also connects the reactor to the separation section thus creating the potential for interactions.

It should be clear from Figs. 5.12 to 5.14 that it would be impossible to build a process so that the flow of heat would balance itself along the three paths for all operating conditions. We could not even bank on this happening for the design conditions because of imperfections in our basic data. We need a control system to help direct the flow of heat just as we need controllers to manage material flow and inventory.

Our first and foremost concern is to dissipate the exothermic heat away from the reactor. This requires that we have the ability to bypass the feed preheaters such that excess heat from the process is not carried back with the reactor feeds. We introduce a set of bypass valves along with temperature controllers as shown in Fig. 5.15. We have elected to install the bypass valves on the cold side to lower the investment cost and facilitate maintenance. We also get better temperature control of the cold stream leaving the preheaters. In Fig. 5.15 we show the stream temperatures at the valve inlets as well as the control points. These data were obtained from the T-H diagram provided by Terrill and Douglas (1987a). Even though we bypass on the cold side, we see that one of the bypass valves has to operate at a temperature over 330°C.

Having secured the integrity of the reactor we next turn to the separation section. The concern here is to avoid picking up too much heat from the hot gas stream traveling through the reboilers. A simple solution to this problem is shown in Fig. 5.16. We have installed four more control valves in the process that bypass the reboilers. The column tray temperature controllers adjust the amount of bypass such that no more heat than necessary is delivered to each column. This design is similar to what was proposed by Terrill and Douglas (1987c).

While the control structure proposed in Fig. 5.16 satisfies the basic demands of heat management, it is not a scheme we would recommend building. First, we have reservations about the design of the recycle column reboiler and its hot bypass. As shown in Fig. 5.16, the hot gas temperature is over 425°C. When we add the fact that the operating pressure is close to 500 psia and that the stream contains mostly methane and hydrogen, we have to wonder how to design the reboiler and its bypass valve so they will operate safely and reliably.

Another concern we have is the propagation of disturbances. Consider, for example, an increase in the quenched reactor effluent temperature, T_1 (Fig. 5.17). We can use Eqs. (5.4) and (5.5) to estimate the effect that this temperature increase has on the streams around the preheater upstream of the furnace. If the bypass flow is small under normal operating conditions, we can assume that $(\dot{m}C_P)_H \approx (\dot{m}C_P)_C \approx$

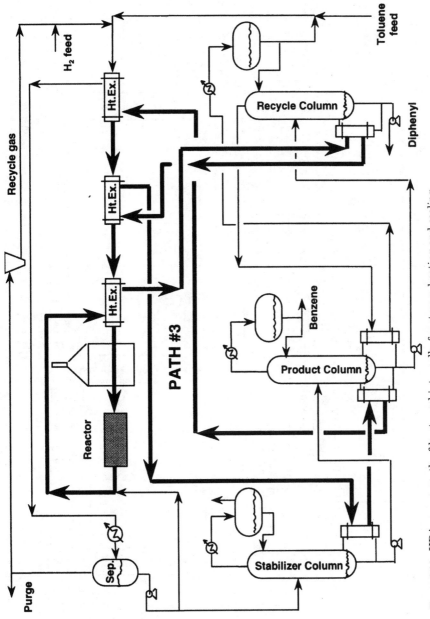

Figure 5.14   HDA process path of heat used internally for stream heating and cooling.

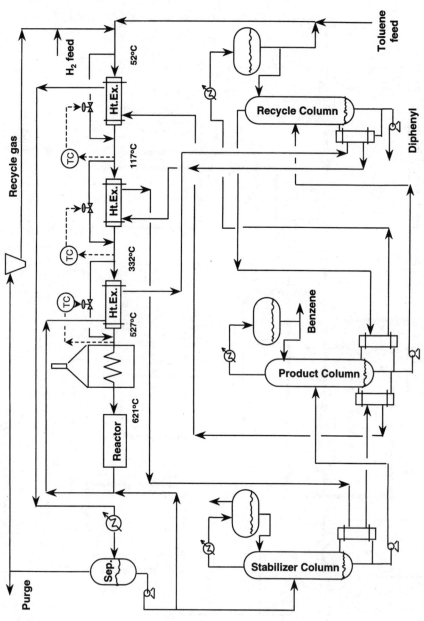

Figure 5.15 HDA process controllers for dissipation of reaction heat by preheater bypass.

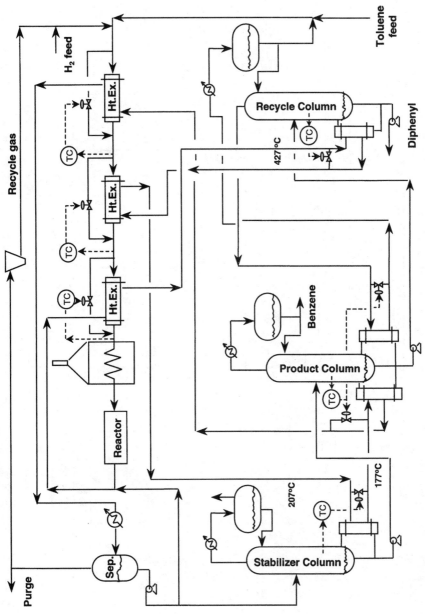

Figure 5.16 HDA process controlling heat for distillation by reboiler bypass.

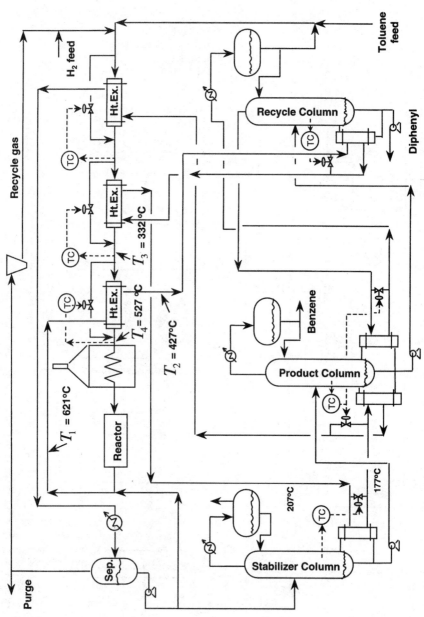

Figure 5.17 HDA process illustration of disturbance propagation in quenched reactor effluent temperature.

$(\dot{m}C_P)_{min}$. The exchanger effectiveness is then obtained from information given in Figs. 5.15 and 5.16.

$$\epsilon = \frac{527 - 332}{621 - 332} = 0.67$$

The exit temperatures are computed from the inlet temperatures according to

$$T_2 = T_1(1 - \epsilon) + \epsilon \cdot T_3$$

$$T_4 = T_3(1 - \epsilon) + \epsilon \cdot T_1$$

We see that for each degree change in the quench temperature T_1, there is a 0.67° change in the furnace feed temperature T_4 and a 0.33° change in the hot stream temperature T_2 going to the recycle column reboiler. However, the preheater bypass controller is controlling T_4 such that ultimately the full change in T_1 will be passed on to T_2. The increased temperature to the recycle column reboiler will increase its boilup. This becomes a disturbance to the column. The column tray temperature controller will respond by increasing the bypass around the reboiler thus deflecting the disturbance to the second preheater.

We can repeat the reasoning logic for the other two preheaters and conclude that the quench temperature upset will migrate to the stabilizer column and the product column before it finally reaches the water cooler. The control system will have done its job as far as heat management is concerned but at the expense of disturbing the entire plant.

The final concern we have about the control structure in Fig. 5.16 is how to start up and turn down the plant. For example, how would we start up the columns without running the furnace and the reactor? Also, how could we turn off the heat to any of the reboilers when the reactor and the furnace are running? The bypass valves may not be designed to take the full gas stream when fully opened. This implies that we need two control valves working in tandem around each reboiler or a three-way valve. Neither of those options is particularly attractive. See Jones and Wilson (1997) for further discussions on process flexibility related to heat integrated designs.

We can solve some of the control difficulties associated with Alternative 6 by adding auxiliary utility coolers and reboilers to the process as suggested by Tyreus and Luyben (1976) and Handogo and Luyben (1987). In Figure 5.18 we show a control configuration that uses three new reboilers and three utility coolers to improve controllability. The coolers are located in bypass streams around the process-to-process reboilers so that disturbances in the heat balance can be dissipated

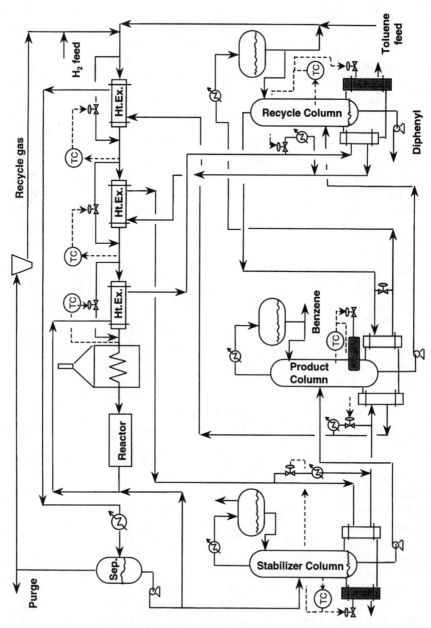

Figure 5.18 HDA process Alternative 6 with complete heat management control system using auxiliary coolers and reboilers.

quickly to utilities without propagating through the entire plant. This is the essence of the rate of exergy destruction principle. Disturbances are rejected to the auxiliary coolers when the column temperature controllers divert excess heat around the main reboilers. The auxiliary reboilers are used to provide a quick source of energy for the columns so that heat deficiencies in the process are not propagated to the next downstream unit operation.

The control system in Fig. 5.18 has a much better chance of providing flexibility and operability than the one shown in Fig. 5.16. However, we have added quite a bit of extra equipment and control valves so the economics of the modified Alternative 6 have to be revisited. In addition, we still have concerns about the design of the hot-gas-heated reboilers and bypass valves for hot hydrogen service. It may be that this level of heat integration is too complicated and ambitious compared to the savings achievable. Simpler designs could be just as profitable when operability is put into the picture. We should recall the 80/20 rule that suggests that we could get 80 percent of the savings with 20 percent of the effort. The base case shown in Chap. 1 and Alternative 1 mentioned earlier represent simpler designs with a healthy level of heat recovery. We therefore take a closer look at the characteristics and control of such alternatives.

5.7 Reactor Feed-Effluent Exchange Systems

5.7.1 Introduction

In Chap. 4 we mentioned that the simplest reactor type from a control viewpoint is the adiabatic plug-flow reactor. It does not suffer from output multiplicity, open-loop instability, or hot-spot sensitivity. Furthermore, it is dominated by the inlet temperature that is easy to control for an isolated unit. The only major issue with this reactor type is the risk of achieving high exit temperatures due to a large adiabatic temperature rise. As we recall from Chap. 4, the adiabatic temperature rise is proportional to the inlet concentration of the reactants and inversely proportional to the heat capacity of the feed stream. We can therefore limit the temperature rise by diluting the reactants with a heat carrier.

While introducing a heat carrier solves the reactor problem, it unfortunately creates some other concerns. First, we increase the size of the separation section since we have to separate the products from a large amount of heat carrier. Second, we make the plant thermally inefficient by significantly increasing the plant's energy load due to repeated heating and cooling of the heat carrier. To solve the efficiency problem we introduce heat integration. As we saw in the HDA example, a healthy improvement in efficiency is achieved by simply preheating the reactor feeds with the hot reactor effluent. These schemes are called *feed-effluent heat exchange* systems, or FEHE systems for short. An FEHE system sometimes has the ability to eliminate all heating and cooling requirements around the reactor and make it autothermal.

It now looks as if we have achieved the best of all worlds: a thermally efficient process with an easy-to-control reactor! Can this be true? Not quite. What we forget are the undesirable effects on the reactor that thermal feedback introduces. In Chap. 4 we explained in detail how process feedback is responsible for the same issues we tried to avoid in the first place by selecting an adiabatic plug-flow reactor. It is necessary that we take a close look at the steady-state and dynamic characteristics of FEHE systems.

5.7.2 Open-loop characteristics

Just as we approached reactor control in Chap. 4, we will start by exploring the open-loop effects of thermal feedback. Consider Fig. 5.19, which shows an adiabatic plug-flow reactor with an FEHE system. We have also included two manipulated variables that will later turn out to be useful to control the reactor. One of these manipulated variables is the heat load to the furnace and the other is the bypass around the preheater. It is clear that the reactor feed temperature is affected by the bypass valve position and the furnace heat load but also by the reactor exit temperature through the heat exchanger. This creates the possibility for multiple steady states. We can visualize the different

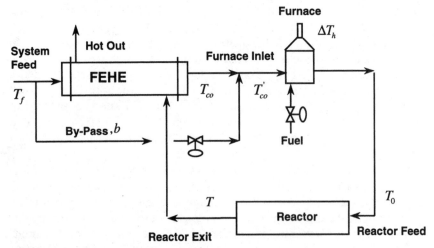

Figure 5.19 Adiabatic plug-flow reactor with feed-effluent heat exchanger and trim heater.

steady states by a graphical technique similar to the one we used for a cooled CSTR in Chap. 4.

We start by plotting the temperature rise in the reactor. This is done by integrating the steady-state differential equations that describe the composition and heat effects as functions of the axial position in the reactor. The adiabatic plug-flow reactor gives a unique exit temperature for a given feed temperature. This also means that we get a unique difference between the exit and feed temperatures. The temperature difference has to be less than or equal to the adiabatic temperature rise at a given, constant feed composition. Figure 5.20 shows the fractional temperature rise as a function of the reactor feed temperature for a typical system.

The next step in the analysis is to seek another functional relationship between the reactor exit temperature and the reactor feed temperature resulting from the heat exchange, bypass, and influence of the furnace. Once we find the second relation we can superimpose it on top of the reactor temperature rise expression shown in Fig. 5.20. Intersections between the two curves constitute open-loop, steady-state solutions to the combined reactor-FEHE system.

Equation (5.5) provides the starting point for our second functional relation. When we use the nomenclature of Fig. 5.19 we get the following simple relationship around the heat exchanger.

$$T_{co} = T_f + \epsilon(T - T_f)$$

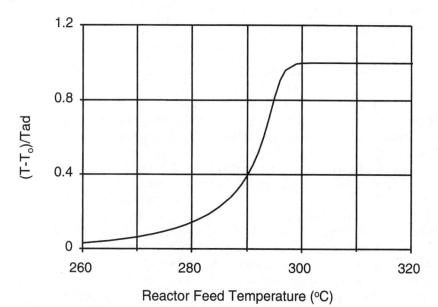

Figure 5.20 Normalized temperature rise in adiabatic plug-flow reactor as function of reactor feed temperature.

This follows since the cold stream to the heat exchanger has the smaller total heat capacity due to the bypass [i.e., $(\dot{m}C_P)_{\min} = (\dot{m}C_P)_C$].

We then add the effect of the bypass as a simple mixing operation:

$$T'_{co} = (1 - b)T_{co} + bT_f$$

where b is the fraction bypass.

The furnace, finally, provides a constant heat input in manual operation. This heat input gives a constant temperature increase ΔT_h. The functional relation between the reactor feed temperature T_0 and the reactor exit temperature T can now be computed:

$$T_0 = (1 - b)[T_f + \epsilon(T - T_f)] + bT_f + \Delta T_h$$

or

$$T = \frac{T_0 + ((1 - b)\epsilon - 1)T_f - \Delta T_h}{(1 - b)\epsilon} \tag{5.10}$$

By subtracting T_0 from both sides of Eq. (5.10) and dividing by the adiabatic temperature rise ΔT_{ad}, we obtain the following relation:

$$\frac{T - T_0}{\Delta T_{ad}} = \left(\frac{1}{(1 - b)\epsilon} - 1\right) \cdot \frac{T_0 - T_f}{\Delta T_{ad}} - \frac{1}{(1 - b)\epsilon}\frac{\Delta T_h}{\Delta T_{ad}} \tag{5.11}$$

Equation (5.11) represents a straight line in the diagram of fractional temperature rise versus reactor feed temperature. We show three such lines in Fig. 5.21. All lines intersect the temperature rise curve at least once (at a low temperature not shown in Fig. 5.21). It therefore appears that the reactor FEHE can have one, two, or three steady-state solutions for this particular set of reaction kinetics. Furthermore, the intermediate steady state, in the case of three solutions, is open-loop unstable due to the slope condition discussed in Chap. 4. This was verified by Douglas et al. (1962) in a control study of a reactor heat exchange system.

Let us now explore in more detail the factors that determine the number of steady-state solutions. First, we notice from Eq. (5.11) that the slope of the line is steep for *small* heat exchangers ($\epsilon \ll 1$) and becomes vertical when there is no heat exchanger at all ($\epsilon = 0$). This is completely opposite to what we observed for a jacketed CSTR in Chap. 4. In a jacketed CSTR the slope of the heat removal line increases with the size of the heat transfer area. The difference is that, in the case of a CSTR, a large heat transfer area increases the rate of heat removal (exergy destruction), driving the system toward stable operation at a single steady state, whereas in the case of an FEHE system,

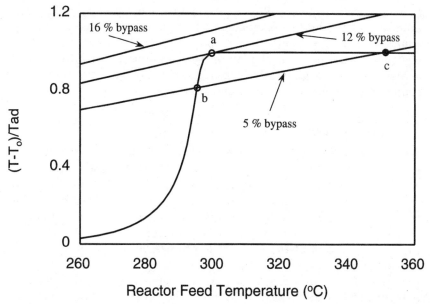

Figure 5.21 Heat removal lines in feed-effluent heat exchange system with bypass.

the large heat transfer area retains heat within the system and thereby promotes the occurrence of multiple solutions.

The second factor affecting the number of steady states is the bypass fraction b. A large bypass fraction is similar to having a small heat exchanger. We say similar because by increasing b we also increase ϵ such that the quantity $(1 - b)\epsilon$ does not change as fast as when ϵ remains constant. The slope of the line still increases with increasing b as shown in Fig. 5.21. The effectiveness and the bypass rate both influence where the straight line intercepts the y axis.

The final factor affecting the number of steady states is the furnace. It affects only the line's intercept and not the slope. A large amount of heat to the furnace lowers the intercept and increases the stable operating temperatures as we might expect. A small heat exchanger and a large furnace give a single, stable steady state.

What are the control implications of this analysis? The first conclusion is that autothermal systems (no furnace) have two or more steady states. There is also a good chance that the normal operating point corresponds to the intermediate steady state that is open-loop unstable. This is certainly the case when the reactor is operated at less than 100 percent conversion.

The trade-off between steady-state economics and controllability should now be obvious. From a steady-state standpoint we would like to have a large heat exchanger for maximum heat recovery and thermal efficiency. This, however, means that we have multiple steady-state solutions and the design case might be open-loop unstable. From a control standpoint we prefer an open-loop stable process that is only achievable with a relatively large, continuously operating furnace. A couple of examples will help illustrate these points.

5.7.3 HDA example

We start with the design and control of the FEHE systems in the HDA process. We show the base case design in Fig. 5.22. The HDA process requires a furnace for all design cases since the reactor effluent stream is quenched down to the reactor feed temperature to prevent by-product formation in the heat exchanger. It is easy to see that when we subtract the effect of the quench, ΔT_q, from the reactor exit temperature T, we raise the intercept of the straight line represented by Eq. (5.11). The furnace must operate to compensate for this effect. The slope of the line is still only affected by the exchanger's heat transfer area (its effectiveness). The interesting feature of the quench is that it provides another manipulated variable that affects the control of the reactor-FEHE system. Since the quench and the furnace have opposing effects on the reactor's feed temperature, we could suspect the possibility of control loop interactions.

The HDA reactor has less than 100 percent per-pass conversion of toluene, meaning that the normal operating point is the intermediate,

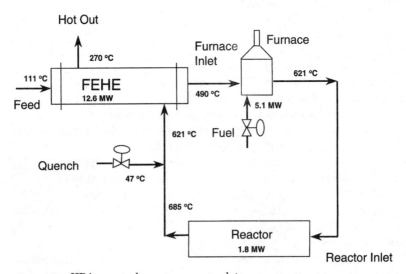

Figure 5.22 HDA process base case process data.

unstable steady state when a sizable heat exchanger is used. We first investigate the control performance of the design case shown in Fig. 5.23. The process is simulated on DuPont's nonlinear dynamic simulator TMODS. We first verify that the system is open-loop unstable by switching the two temperature controllers into manual mode and perturbing the system slightly. The reactor either quenches or becomes very hot, depending upon the direction of the perturbation. Next, we tune the quench temperature controller. This is easy to do since the dynamics of the mixing process are fast and there are no other significant delays in the loop. The interesting aspect of putting just the quench temperature controller in automatic while leaving the furnace in manual is that the system is stabilized. When the quench temperature is controlled, the reactor feed temperature is indirectly controlled as well. Another way of looking at it is to say that the gain between the reactor exit and the reactor feed is reduced. This lowers the overall loop gain to less than one in the positive feedback loop formed by the reactor, heat exchanger, and the furnace. The stabilizing effects of partial control were discussed in Chap. 4 and are further addressed by Silverstein and Shinnar (1982) in relation to reactor-FEHE systems. With the reactor system stabilized it is trivial to tune the reactor feed temperature controller by use of a relay test.

CS1 - Big Furnace/Small Heat Exchanger

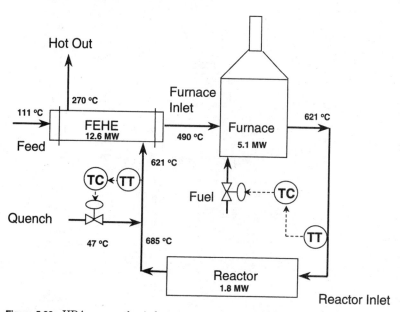

Figure 5.23 HDA process basic heat management control system.

We also tried to stabilize the process by tuning the reactor feed temperature loop while leaving the quench loop in manual. This turned out to be difficult to do since it requires a trial-and-error tuning approach for the open-loop unstable system. However, it was possible to find a combination of gain and reset time that stabilized the system even though it ended up very underdamped. The reason for the difficulties here as opposed to the quench loop is that we assume that the furnace has a much longer time constant than the thermal mixing in the quench loop.

We next try a more aggressive heat recovery alternative as shown in Fig. 5.24. The heat input to the furnace is quite small and most of the heat is provided by the large feed-effluent exchanger. With our choice of measurement lags (two 1-minute lags in series) and the lag in the furnace, this system cannot be stabilized by feedback control around the furnace if the quench controller is in manual. However, it is possible to stabilize the system with just the quench controller in automatic and the furnace controller in manual. Subsequent tuning of the furnace controller is then easy since the new system is open-loop stable.

As we have seen several times in this chapter, a bypass around the heat exchanger introduces another manipulated variable for control. We show such a design in Fig. 5.25. The bypass ought to be as fast as the quench and it should be possible to stabilize the system with just

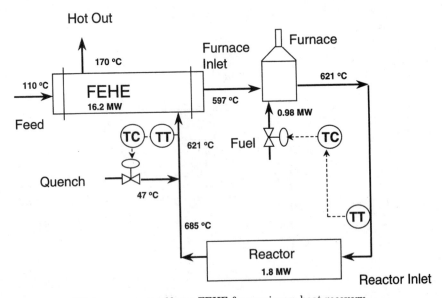

Figure 5.24 HDA process use of large FEHE for maximum heat recovery.

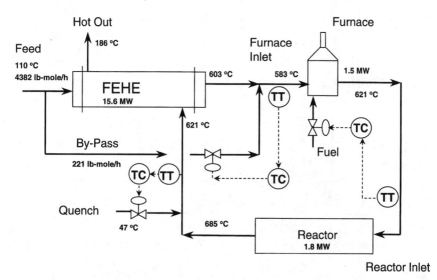

Figure 5.25 HDA process with large FEHE and bypass control.

the bypass loop in automatic. Indeed, we found this to be true. This is a great example of the meaning of partial control discussed in Chap. 4. For the HDA system we happen to have only one dominant variable (reactor inlet temperature) but three manipulated variables. It is sufficient to use only one fast manipulated variable that affects the dominant variable to have unit control. The other manipulators can then be used to extend the range of stable operation.

All three design cases discussed so far for the HDA process have a finite tolerance to disturbances, since the normal operating point is open-loop unstable. In other words, it is always possible to disturb either of the systems enough to throw the entire process into wild oscillations that extend through the two open-loop stable steady states. The tolerance to such disturbances improves as the size of the furnace increases and as we introduce more flexibility in terms of additional manipulated variables like bypasses and quench loops. However, these extra manipulated variables can never remove the existence of the unstable operating point like a large furnace can. The most stable design, but also the most expensive, is the one with a large furnace and no feed-effluent exchanger.

We conclude this discussion by showing the closed-loop response to two different disturbances. In Fig. 5.26 we reduce the setpoint of the reactor feed temperature controller. In Fig. 5.27 we reduce the amount of toluene fed to the reactor. The changes shown were the largest that could be handled by the system with the small furnace and the large exchanger without a bypass. The design with the bypass (CS2) and the

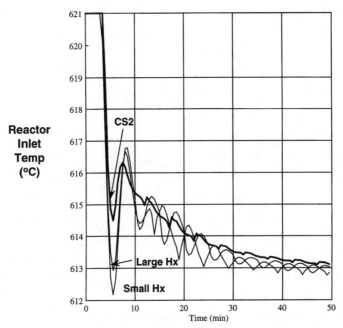

Figure 5.26 Dynamic response of HDA reactor inlet temperature to −8°C setpoint change for three different process and control configurations.

design with the large furnace handle these disturbances quite well and they also tolerate larger upsets than the ones shown here without going unstable.

5.7.4 Reactors with wrong-way behavior

The HDA reactor is unpacked and therefore cannot exhibit the wrong-way behavior discussed in Chap. 4. However, it is quite common that gas phase reactions are carried out over a catalyst so it is important to understand the implications of the wrong-way behavior on the control of reactors with feed-effluent heat exchangers.

To that end we have constructed a simulation of a fictitious system that has a severe inverse response. We show the design in Fig. 5.28 and give the design parameters in Table 5.1. The reactor has a large Lewis number (Le = 25), nearly complete per pass conversion of the reactant, and little axial dispersion. These are all factors necessary for wrong-way behavior. In fact the example plot of wrong-way behavior shown in Chap. 4 was generated from this reactor.

We start by exploring the open-loop characteristics of the autothermal system with a 12 percent bypass rate. We already showed the steady-

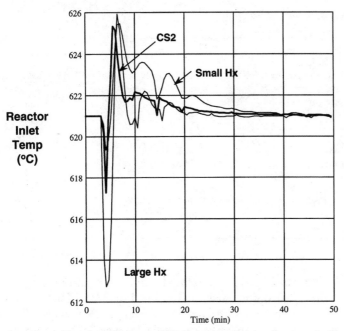

Figure 5.27 Dynamic response of HDA reactor inlet temperature to 15 percent decrease in toluene recycle rate for three different process and control configurations.

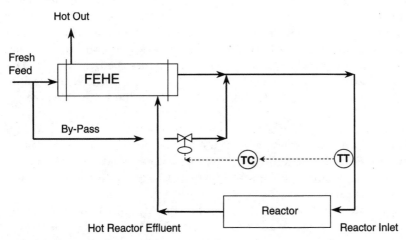

Figure 5.28 Control of packed adiabatic plug-flow reactor with FEHE and bypass control.

TABLE 5.1   Design Data for Reactor with
Inverse Response

Molecular weight component A	100
Molecular weight component B	100
Vapor heat capacity component A	20 kcal/kmol·K
Vapor heat capacity component B	20 kcal/kmol·K
Production rate of component B	100 kmol/h
Operating pressure	15 atm
Reactor design exit temperature T	380°C
Activation energy E_a/R	20,000 K
Heat of reaction ΔH	−17.6 kcal/mol
Reactor feed flowrate	1100 kmol/h
Mole fraction A in reactor feed	0.09
Reactor feed temperature T_0	300°C
Reactor void (vapor) space	123 m³
Packing weight	123,000 kg
Packing heat capacity	0.15 kcal/kg·°C

state operating point in Fig. 5.21. The heat exchanger line is tangent to the reactor temperature rise curve (point a in Fig. 5.21). This means that the hot and the intermediate steady states have merged. The combined steady state is dynamically unstable. We verify this by simulating the system shown in Fig. 5.28 with a fixed bypass rate. The results are shown in Fig. 5.29. The bypass valve is held constant from time zero. After about 5 hours of operation, in this mode, the feed and exit temperatures start oscillating. The oscillations grow until the reactor temperatures enter fixed-amplitude-limit cycles.

We next verify that there exists a hot stable steady state when the system equations have three stationary solutions. With the temperature controller in manual, we lower the bypass rate from 12 percent to 5 percent. According to Fig. 5.21 this should create three steady states and we would expect the intermediate state to be unstable. Figure 5.30 shows what happens when we reduce the bypass rate. The reactor feed temperature goes up initially due to reduced bypassing. This makes the reactor exit temperature drop due to the inverse response. The drop in the reactor exit temperature causes a drop in the feed temperature below the intermediate steady state (point b in Fig. 5.21). However, this state is unstable so the temperatures continue to oscillate. Slowly, but surely, the temperatures trend toward the hot steady state (point c in Fig. 5.21) where the oscillations die out and the reactor remains stable.

In a third experiment we attempted to extinguish the reactor by increasing the bypass to 16 percent. According to Fig. 5.21 this creates a single stable steady state (cold with little or no conversion). However, the result of this experiment, shown in Fig. 5.31, was surprising. Instead of migrating to the cold steady state, the whole system enters into

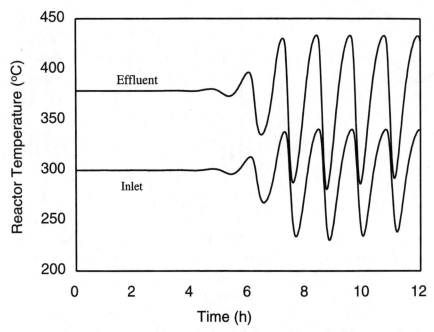

Figure 5.29 Open-loop dynamic behavior of packed adiabatic plug-flow reactor with FEHE.

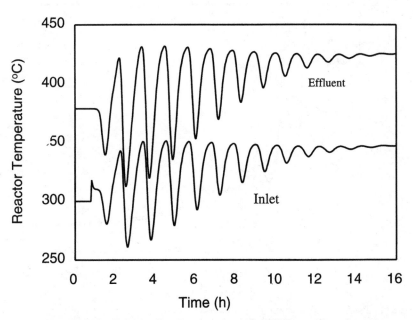

Figure 5.30 Packed adiabatic plug-flow reactor with FEHE open-loop dynamic response to decrease in bypass flow from 12 to 5 percent.

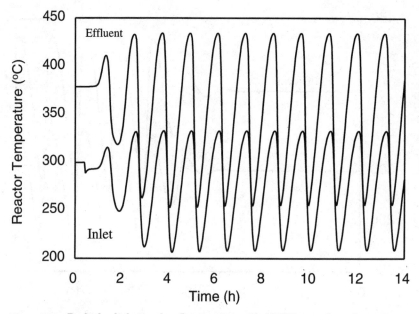

Figure 5.31 Packed adiabatic plug-flow reactor with FEHE open-loop dynamic response to increase in bypass flow from 12 to 16 percent.

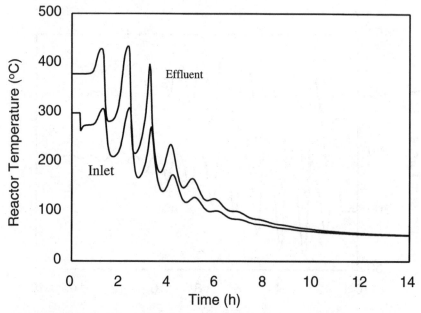

Figure 5.32 Packed adiabatic plug-flow reactor with FEHE open-loop dynamic response to increase in bypass flow from 12 to 20 percent.

a large-amplitude limit cycle. This is one of the *dynamic* implications of nonlinearities and positive feedback. When we open the bypass further the limit cycle cannot be sustained and the system cools down as we would expect. We show this in Fig. 5.32, where the bypass rate is increased to 20 percent.

After demonstrating these serious dynamic problems for the open-loop reactor exchange system, we might suspect that it could be difficult to stabilize such a reactor around one of the unstable steady states. This suspicion is correct, as was demonstrated in a study by Tyreus and Luyben (1993). Unexpected results were obtained for the tuning of a PI controller to hold the reactor feed temperature by manipulating the bypass. In particular, it was found that the temperature loop was stabilized with small controller reset times. This is counterintuitive, since integral action is expected to reduce stability because of the increase in phase shift. This is true at high frequencies. At low frequencies, where the inverse response and the deadtime dominate the response, the high loop gain provided by the integral action is required to stabilize the otherwise unstable process.

5.7.5 Summary

We have discussed in detail the design, open-loop behavior, and control of adiabatic, plug-flow reactors with feed-effluent exchangers. These systems are attractive from a steady-state economic viewpoint but can exhibit serious operational concerns. It is necessary to make a careful evaluation of these systems to ensure that the steady-state savings are not offset by control problems. In particular, packed reactors with a severe inverse response are difficult to control when coupled with a heat exchanger. The control problems are best tackled by process design changes. Empty reactors, large furnaces, heat exchanger bypasses, and quench systems all promote stable operation.

5.8 Conclusion

In this chapter we have discussed the management of energy in a chemical process. We showed that chemical processes are generally thermally inefficient and that there is a strong incentive to improve the efficiency by heat integration. This is done with heat exchangers of various kinds, and we discussed possible control schemes. We also took a strategic view of the heat management process and identified different heat pathways. This provided a rationale for implementing the plantwide control structure. The essence of the approach is to provide auxiliary utility exchangers to prevent thermal disturbances from propagating through the plant. Since these auxiliary exchangers can add significantly to the cost of the plant, we indicated that more cost-effective designs merely involve reactor feed-effluent heat exchangers. The dynamics and control of these systems were discussed in detail. We found that the most stable systems include a large furnace and bypasses around the feed-effluent exchangers.

5.9 References

Douglas, J. M., Orcutt, J. C., and Berthiaume, P. W. "Design and Control of Feed-Effluent, Exchanger-Reactor Systems," *Ind. Eng. Chem. Fundam.*, **1**, 253–257 (1962).

Douglas, J. M. *Conceptual Design of Chemical Processes*, New York: McGraw-Hill (1988).

Gebhart, B. *Heat Transfer*, 2d ed., New York: McGraw-Hill (1971).

Handogo, R., and Luyben, W. L. "Design and Control of a Heat-Integrated Reactor/ Column Process," *Ind. Eng. Chem. Res.*, **26**, 531–538 (1987).

Jones, W. E., and Wilson, J. A. "An Introduction to Process Flexibility. 1. Heat Exchange," *Chemical Engineering Education*, **31**, 172–177, Summer (1997).

Silverstein, J. L., and Shinnar, R. "Effect of Design on the Stability and Control of Fixed Bed Catalytic Reactors with Heat Feedback. 1. Concepts," *Ind. Eng. Chem. Proc. Des. Dev.*, **21**, 241–256 (1982).

Terrill, D. L., and Douglas, J. M. "A *T-H* Method for Heat Exchanger Network Synthesis," *Ind. Eng. Chem. Res.*, **26**, 175–179 (1987a).

Terrill, D. L., and Douglas, J. M. "Heat-Exchanger Network Analysis. 1. Optimization," *Ind. Eng. Chem. Res.*, **26**, 685–691 (1987b).

Terrill, D. L., and Douglas, J. M. "Heat-Exchanger Network Analysis. 2. Steady-State Operability Evaluation," *Ind. Eng. Chem. Res.*, **26**, 691–696 (1987c).

Tyreus, B. D., and Luyben, W. L. "Controlling Heat Integrated Distillation Columns," *Chem. Eng. Prog.*, **72**, No. 9, 59–66 (1976).

Tyreus, B. D., and Luyben, W. L. "Unusual Dynamics of a Reactor/Preheater Process with Deadtime, Inverse Response and Openloop Instability," *J. Proc. Cont.*, **3**, 241–251, (1993).

6

Distillation Columns

6.1 Introduction

More work has appeared in the chemical engineering literature on distillation column control than on any other unit operation. Books on this important subject date back to the pioneering work of Rademaker et al. (1975), Shinskey (1977), and Buckley et al. (1985). Some of the more recent developments are discussed in Luyben (1992). The long-term popularity of distillation control is clear evidence that this is a very important and challenging area of process control. Most chemical plants and all petroleum refineries use distillation columns to separate chemical components. Distillation is the undisputed king of the separation processes.

We cannot hope to cover the vast subject of distillation fundamentals and control in a single chapter in this book. Our objective here is to review some of the basic principles about distillation and then to summarize the essentials of distillation column control, particularly as it relates to the plantwide control problem. Many more details are available in the books cited above. In the first section we review some of the important fundamentals about distillation. Sections 6.3 through 6.8 discuss distillation column control mostly from the perspective of an isolated column or column system. This treatment presents what may appear to be a "laundry list" of control structures for different types of columns. Although we feel this presentation is valuable, particularly for the young inexperienced student or engineer, it is also vital to retain a broader plantwide perspective. Section 6.9 addresses some of these plantwide distillation control issues. In the next section we review the process fundamentals of distillation as they relate to process operation and control.

6.2 Distillation Fundamentals

6.2.1 Vapor-liquid equilibrium

We recall from thermodynamics that a component's contribution to a mixture's ability to perform mechanical work (the Gibbs free energy) is called the *chemical potential* μ_i. The chemical potential increases with temperature, pressure, and concentration of the component in the mixture. For example, the chemical potential for an ideal gas component can be expressed as

$$\mu_i = \mu_i^0 (T, P_0) + RT \ln \frac{P}{P_0} + RT \ln y_i \qquad (6.1)$$

where $\mu_i^0 (T, P_0)$ = chemical potential of pure, gaseous
component at temperature T and
reference pressure P_0
R = gas law constant
P = mixture pressure
y_i = mole fraction of component i in mixture

The conditions for phase equilibrium dictate that the chemical potential for a given component is the same in all phases. Distillation is a separation method that takes advantage of this fact and the chemical potential's dependency on pressure and composition. To see how this works we consider a vapor-liquid equilibrium system at pressure P, temperature T, and mole fractions x_i and y_i in the liquid and vapor phases respectively. Equation (6.1) expresses the chemical potential of component i in the vapor phase, provided the system pressure is not too high.

To find an equivalent expression for μ_i in the liquid phase we start with a pure component in the gas phase at pressure P_0 and temperature T. The chemical potential is $\mu_i^0(T,P_0)$. We then compress the gaseous component until we reach the saturation pressure and the gas liquefies. This occurs at P_i^s, which is the same as the pure component vapor pressure at temperature T. Provided that the gas behaves close to ideally during the compression, the chemical potential increases by $RT \ln (P_i^s/P_0)$ energy units during this step. The pressure of the pure component liquid is then adjusted to the system pressure P. The change in the chemical potential from this action is usually small and will be ignored. Finally, we mix the pure liquid component with the other components so its mole fraction becomes x_i. The dilution is a spontaneous process (entropy increasing) that reduces the chemical potential by $RT \ln x_i$ energy units, provided all the components mix ideally.

The final potential of component i in the liquid mixture is therefore

$$\mu_i^L = \mu_i^0 \, (T, P_0) + RT \ln \frac{P_i^s}{P_0} + RT \ln x_i \tag{6.2}$$

Since the chemical potentials have to be equal between the vapor and the liquid phases, we obtain from Eqs. (6.1) and (6.2) the familiar Raoult's law:

$$Py_i = P_i^s x_i \tag{6.3}$$

We now see why the vapor composition has to be greater than the liquid composition for components that have a vapor pressure greater than the system pressure (i.e., $P_i^s > P$). The reverse is true for components with vapor pressures lower than the system pressure.

Equations (6.1) and (6.2) pertain to ideal systems, that is, systems where there are no interactions between the molecules. In a real system the pressure effect on μ_i in the vapor phase has to be modified by a fugacity coefficient Φ_i, and the effect of mixing on the chemical potential in the liquid phase has to be modified by an activity coefficient γ_i. The more general expression for equilibrium (called the Φ-γ representation) then becomes

$$P\Phi_i y_i = P_i^s \gamma_i x_i \tag{6.4}$$

The fugacity coefficient is usually obtained by solving an equation of state (e.g., Peng-Robinson, Redlich-Kwong). The activity coefficient is obtained from a liquid phase activity model such as Wilson or NRTL (see Walas, 1985).

In solving distillation problems, it is often convenient to express the vapor composition as a function of the liquid composition. The ratio between the two compositions is the K value,

$$\frac{y_i}{x_i} \equiv K_i(T,P,x,y) = \frac{P_i^s \gamma_i}{P\Phi_i} \tag{6.5}$$

Unfortunately, K values are generally composition-dependent through the fugacity and activity coefficients. Only in ideal systems is the composition dependency removed:

$$\frac{y_i}{x_i} = K_i(T,P) = \frac{P_i^s(T)}{P} \tag{6.6}$$

However, the ideal system K values still depend upon pressure and temperature. Since the Gibbs phase rule dictates that only one of those two intensive variables can be set (in addition to the liquid phase mole fractions for a two-phase system), we still have to solve for the other variable before the K values can be evaluated. To eliminate the implicit dependency on temperature, we can introduce relative volatilities α_{ij} that are ratios of the K values:

$$\alpha_{ij} \equiv \frac{K_i}{K_j} \tag{6.7}$$

For ideal systems (adhering to Raoult's law) the relative volatility reduces to the ratio between two components' vapor pressures. The dependency on total pressure is eliminated, and there is only a weak residual dependency on temperature. In many systems the relative volatilities can be considered constant, and this provides an easy means of calculating the vapor phase composition given the liquid phase mole fraction:

$$y_i = \frac{\alpha_{iR} \cdot x_i}{\sum\limits_{n} \alpha_{iR} \cdot x_i} \tag{6.8}$$

where n = number of components in mixture and R = reference component with relative volatility equal to unity.

It is interesting to note that the vapor and liquid compositions are usually different for ideal mixtures. We can see this from Eq. (6.6), since different pure component vapor pressures are rarely equal at the same temperature. This picture changes when nonideal mixtures are considered. As we see from Eq. (6.5), the vapor and liquid mole fractions can become equal when the fugacity and activity coefficients alter the pressure ratio enough to cause the K value to become unity. We then have an *azeotrope*.

Azeotropes can boil at temperatures that are higher than the boiling point temperature of the heavy pure component (maximum-boiling azeotrope) or can boil at temperatures that are lower than the boiling point temperature of the light pure component (minimum-boiling azeotrope). Maximum-boiling azeotropes occur when there is attraction between the components, and minimum-boiling ones occur when there is repulsion.

Azeotropes can also be homogeneous (single liquid phase) or heterogeneous (multiple liquid phases). In a heterogeneous azeotrope the repulsive forces between different molecules in the liquid phase are strong enough to overcome the entropy increase due to mixing such that the liquid splits into two or more separate liquid phases. At equilibrium the chemical potential for each component is still the same in all phases:

$$\mu_i^I = \mu_i^{II} = \ldots = \mu_i^V \tag{6.9}$$

Equality of the chemical potentials in two *liquid* phases translates into a particularly simple relation,

$$\gamma_i^I x_i^I = \gamma_i^{II} x_i^{II} \tag{6.10}$$

where μ_i^I = chemical potential of component i in phase I and μ_i^{II} = chemical potential of component i in phase II. The activity coefficients can force the liquid phase compositions to be quite different. This effect is sometimes used to help separate azeotropes.

6.2.2 Residue curve maps

An ideal mixture of n components requires a sequence of $n - 1$ conventional distillation columns (two product streams) to separate the components completely. The columns can be arranged sequentially without recycle between them. This picture changes when mixtures forming azeotropes must be separated. Nonideal systems sometimes require complex distillation arrangements involving more than $n - 1$ columns with recycle of material between the columns. For the analysis of such systems, we recommend the use of *residue curve maps*. We base the following summary on the excellent book by Doherty and Malone (1998), who pioneered the use of these techniques.

A residue curve is the result of a simple batch experiment run at constant pressure. Consider a flask containing a liquid mixture with initial composition $x_1^0, x_2^0, \ldots, x_n^0$. We heat the flask such that the mixture boils. As vapors are distilled from the flask, the composition (and equilibrium temperature) of the remaining liquid (the residue) changes. The composition change can be modeled with n ordinary differential equations:

$$\frac{dM}{dt} = -V \tag{6.11}$$

$$\frac{d(Mx_i)}{dt} = -Vy_i \qquad i = 1, 2, \ldots, n - 1 \tag{6.12}$$

where M = molar holdup in flask and V = molar vapor rate leaving flask. These equations are simplified to give

$$\frac{dx_i}{d\zeta} = x_i - y_i \qquad i = 1, 2, \ldots, n - 1 \tag{6.13}$$

where ζ is dimensionless time $t(V/M)$.

The system of Eq. (6.13) can be integrated forward and backward in dimensionless time starting from the initial condition $x_1^0, x_2^0, \ldots, x_n^0$. The result of the integration gives the time evolution of all the residue

Water

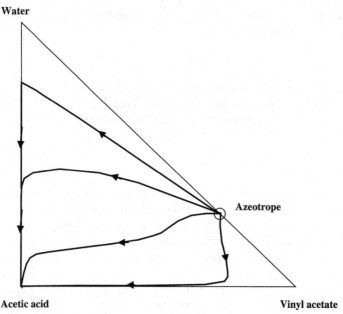

Acetic acid   Vinyl acetate

Figure 6.1   Residue curve map for vinyl acetate, water, and acetic acid.

compositions. Instead of plotting these compositions against time we can plot them against each other to get what is called a residue curve. Fig. 6.1 shows several residue curves for the ternary system vinyl acetate/water/acetic acid. In this system the highest boiler is acetic acid and the lowest boiler is the azeotrope between vinyl acetate and water. We see that all the residue curves start at the low boiling azeotrope and terminate in the high boiler. The complete picture of all possible residue curves is called a *residue curve map*.

The significance of the residue curve map is that it represents the possible composition profiles that a continuous distillation tower would show when operating under total reflux. The map immediately shows what product compositions are feasible given certain feed compositions. In some systems the map contains several distillation regions separated by distillation boundaries. It is not possible to leave a certain region by simple distillation. The residue curve map shown in Fig. 6.1 has only one region. Any feed to this separation system can be separated into pure acetic acid and the vinyl acetate/water azeotrope, given enough theoretical stages.

While the residue curves pertain exactly to total reflux conditions, they can also be used to estimate the behavior of a column with a continuous feed. To help analyze these situations we make use of the

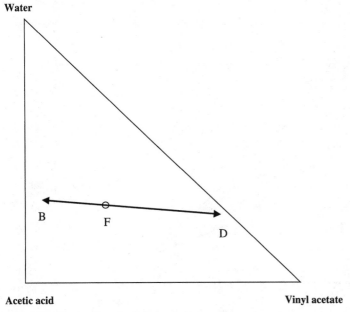

Figure 6.2 Illustration of mixing rule in ternary system.

three-component mixing rule. The rule is illustrated in Fig. 6.2. Here we show a liquid feed F that is separated into a liquid distillate D and a bottoms stream B. The mixing rule dictates that the three stream composition points have to lie on a straight line in the composition diagram. Furthermore, the distance between the product points and the feed point is inversely proportional to the amount of product taken. For example, in Fig. 6.2 we are taking roughly twice as much bottoms product as we are producing distillate. With the feed composition indicated in Fig. 6.2, the bottoms product must necessarily contain a fair amount of water.

Another feature of a residue map we would like to illustrate is the representation of systems that form two liquid phases. In Fig. 6.3 we show how mixtures of vinyl acetate and water form two liquid phases with drastically different compositions. We can take advantage of this nonideality to help produce pure acetic acid from a single distillation column. In Fig. 6.4 we show how the net feed to a column can be changed by mixing the original feed with the vinyl acetate rich reflux. The new feed composition contains less acetic acid acid and water and more vinyl acetate. When we look at the residue curves that pertain to the new feed composition, we find that they move over areas with little water. Most of the feed water is rejected with the overhead vapors

Water

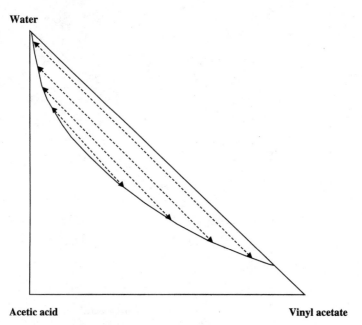

Acetic acid   **Vinyl acetate**

Figure 6.3 Liquid-liquid solubility region.

Water

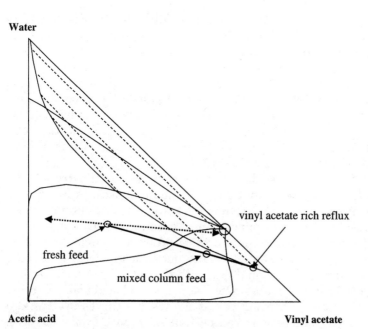

vinyl acetate rich reflux

fresh feed

mixed column feed

Acetic acid   **Vinyl acetate**

Figure 6.4 Illustration of azeotropic distillation on residue curve map.

and never allowed to enter again due to the phase splitting that takes place in the overhead decanter.

6.2.3 Energy requirements

The processes of chemical reaction and heat transfer discussed in the previous two chapters occurred spontaneously without the need for external energy. A separation process like distillation, on the other hand, is not spontaneous and therefore requires energy to proceed. Work is needed to raise the chemical potential of the components in the product streams over their values in the mixed feed. From this viewpoint it might be tempting to assume that the net energy requirement for distillation is the change in the Gibbs free energy between the products and the feed. This is, however, only approximately true since the products are normally taken out at different temperatures than the feed temperature. Instead, as we show in Appendix A, the correct way of evaluating the minimum work requirement for a distillation column is to examine the change in exergy between the products and the feed:

$$\dot{w}_{sep} = \Delta B_f = B_{fD} + B_{fB} - B_{fF}$$
$$= m_D(h_D - T_0 s_D) + m_B(h_B - T_0 s_B) - m_F(h_F - T_0 s_F) \quad (6.14)$$

where $\dot{w}_{sep}$ = power requirement for separation
        B_f = exergy flowrate
        m = molar flowrate
        h = specific enthalpy
        s = specific entropy
        T_0 = absolute temperature of environment

Equation (6.14) is independent of the method of delivering the required work. In most systems the work is derived from condensing steam at a high temperature in the reboiler and then removing heat to ambient conditions through cooling water in the condenser. For such columns we can express the work of separation in terms of the heat input and removal rates by applying a steady-state energy balance to Eq. (6.14):

$$\dot{w}_{sep} = m_D(h_D - T_0 s_D) + m_B(h_B - T_0 s_B) - m_F(h_F - T_0 s_F)$$

$$= \Delta \dot{H} - T_0 \Delta \dot{S} + \dot{q}_{reb} - \dot{q}_{reb} + \dot{q}_{cond} - \dot{q}_{cond}$$

$$+ T_0 \frac{\dot{q}_{reb}}{T_{reb}} - T_0 \frac{\dot{q}_{reb}}{T_{reb}}$$

$$= \dot{q}_{reb}(1 - \frac{T_0}{T_{reb}}) - T_0 \dot{\sigma} \quad (6.15)$$

where $\Delta \dot{H}$ = net enthalpy flow in streams around column
$\Delta \dot{S}$ = net entropy flow in streams
$\dot{q}_{reb}$ = heat flow *supplied* to reboiler
T_{reb} = process-side reboiler temperature
$\dot{q}_{cond}$ = heat flow *removed* in condenser
$T_0 \dot{\sigma}$ = lost work production rate as a result of operating process

Equation (6.15) tells us that the work required for the separation is derived from the available work in the heat to the reboiler after all lost work has been subtracted. The lost work comes from irreversibilities in the operation such as from pressure drop and from mixing vapor and liquid streams that have different compositions and temperatures (see Ognisty, 1995). It is interesting to note that by reducing the lost work contribution we can lower the heat requirement to the reboiler, $\dot{q}_{reb}$, while keeping the same reboiler temperature T_{reb}, (e.g., less reflux with more trays). Or we can lower the average reboiler temperature T_{reb} by the use of intermediate reboilers, and maintain the same net heat input $\dot{q}_{reb}$ (but lower exergy input). A third design alternative is to reduce the heat input to the reboiler and raise the reboiler temperature while still removing heat at T_0. This is what is done with heat-integrated columns.

From a control standpoint we seek strategies and designs that allow us to *alter* quickly the exergy destruction rate $T_0 \Delta \dot{\sigma}$. The total rate of entropy production is

$$\dot{\sigma} = \frac{\dot{q}_{cond}}{T_0} - \frac{\dot{q}_{reb}}{T_{reb}} + \Delta \dot{S}$$

$$\approx \dot{q}_{reb} \left(\frac{1}{T_0} - \frac{1}{T_{reb}} \right) + \Delta \dot{S}$$

When the heat input to the reboiler is used as a manipulated variable for control, we get the fastest response for those designs where the heat is delivered over a large temperature gradient:

$$\frac{T_0 \Delta \dot{\sigma}}{\Delta \dot{q}_{reb}} = \frac{T_{reb} - T_0}{T_{reb}}$$

This has implications in two different cases. For columns with auxiliary reboilers, we should obtain better control by using heat input to the base reboiler (operating at the highest temperature) compared with heat input to the other reboilers. We should also expect better control for columns with large temperature differences between the top and bottom than for columns with small temperature differences.

6.2.4 Reactive distillation

This book deals with control strategies for integrated plants. In most cases the plant consists of a separate reaction section followed by a refining section with recycles between the two. Optimal designs usually call for tight integration of the unit operations, as we mentioned in Chaps. 2 and 5. What if we could take the ultimate step in integration and combine the reaction and separation sections? We would then have *reactive distillation*.

However, we have noticed in previous chapters that the more coupled the process is, the more potential difficulties we have with operation and control. The reason has to do with snowballing, trapping of components, and propagation of composition and thermal disturbances. These issues are definitely still present in reactive distillation, making the control system design for these systems a challenging problem.

Reactive distillation is certainly not for all processes. First, the reaction should occur in the liquid phase to achieve reasonable concentrations and rates. Second, the reaction temperature must be in the range of distillation temperatures. Finally, a suitable catalyst that can operate in a distillation column must be available.

Systems that have the most potential for reactive distillation are those where the reaction is reversible, heat of reaction is not excessively large, and the products have the correct volatilities in relation to the reactants. Those systems reach chemical equilibrium (i.e., reaction stops) unless the reactants are in large excess or the products are continuously removed. An example system has been reported in the literature by Eastman Chemical (Agreda et al., 1990) for the production of methyl acetate from methanol and acetic acid. The discussion about process operation and the control strategy shown in the paper certainly adhere to the plantwide control principles we have outlined in this book.

We do not discuss the control of reactive distillation columns in this book. Although reactive distillation has been used for many years in specific industrial applications, only recently have systematic studies appeared in the open literature on both steady-state design and dynamic control (Doherty and Buzad, 1992; Sneesby et al., 1997). Because generic understanding of this technology is still in an early stage of development, we feel it would be premature to speculate on specific recommendations. However, we are confident that many of the ideas and techniques discussed in this book will apply directly to this microcosm of the plantwide control problem.

6.2.5 Open-loop behavior

In Chaps. 4 and 5 we discussed the open-loop behavior of reactors with process feedback. We showed how nonlinearities combined with feedback can result in multiple steady states. We also showed that when there is a source of production in the feedback loop of a variable (e.g., temperature or autocatalytic component) that promotes further production of the same variable, the resulting amplification can lead to instablity.

Distillation columns have process feedback through the reflux flow from the condenser and through the vapor from the reboiler. However, most conventional distillation systems are not nonlinear enough give rise to multiple steady states. This situation changes when we consider highly nonideal systems or reactive distillation. For these designs it becomes possible to have output multiplicity as we discussed in Chap. 4. In those cases where the desired steady state is unstable, we must use feedback control to stabilize the column. Fortunately, tray temperatures are often dominant variables in distillation systems and the separation is always affected by the net energy input. This makes it fairly easy to find a suitable control loop that will stabilize the unstable operating point.

So far we have been discussing output multiplicity as the primary source of open-loop instability. It is also possible to have input multiplicity in distillation systems. Input multiplicity means that we can get the same output for different levels of the input variables.

Another phenomenon of highly nonlinear systems is parametric sensitivity. We illustrated this behavior for the temperature profile in the plug-flow reactor. Nonideal distillation systems can also show this sensitivity. For example, in Fig. 6.5 a small change in the feed composition or organic reflux flow can dramatically change the composition (and temperature) profile in the column. Instead of a vinyl acetate–rich profile in the top section, a water-rich profile can be present.

6.3 Control Fundamentals

In this section we discuss some basic concepts concerning distillation control: degrees of freedom, basic manipulated variables, and constraints.

6.3.1 Control degrees of freedom

Figure 6.6 shows a simple two-product distillation column and gives the notation we use for flowrates, compositions, and tray numbering. Feed is introduced on tray N_F, numbering from the bottom. There are N_T trays in the column. The molar feed flowrate is F, its composition is z_j (mole fraction of component j), and its thermal condition is q (saturated liquid is $q = 1$, saturated vapor is $q = 0$). The heat transfer rates are Q_R in the reboiler and Q_C in the condenser. Distillate product

Water

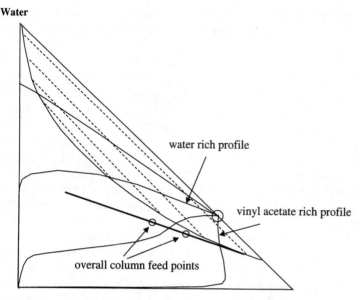

Figure 6.5 Illustration of parametric sensitivity in azeotropic distillation.

is produced at a molar flowrate D with composition $x_{D,j}$. Bottoms product is produced at a molar flowrate B with composition $x_{B,j}$. Reflux and vapor boilup molar flowrates are R and V, and the reflux ratio is $RR = R/D$.

There are six control valves associated with the column, therefore

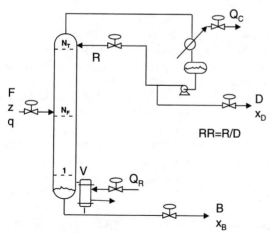

Figure 6.6 Basic distillation column.

there are six control degrees of freedom. One of these is used to set throughput. This typically is the valve in the feed line, but in some columns it is one of the product valves (when this product is an *on-demand* stream). Sometimes throughput is set by the valve on the cooling water to the condenser (if capacity is limited by maximum pressure/maximum cooling conditions) or by the valve on the steam to the reboiler (if capacity is limited by heat input or column flooding).

Two of the control degrees of freedom must be consumed to control the two liquid levels in the process: reflux drum level and base level. Reflux drum level can be held by changing the flowrate of the distillate, the reflux, the vapor boilup, the condenser cooling, or the feed (if the feed is partially vapor). Each of these flows has a direct impact on reflux drum level. The most common selection is to use distillate to control reflux drum level, except in high reflux-ratio columns (RR > 4) where "Richardson's rule" suggests that reflux should be used.

Column base level (or reboiler level in a kettle reboiler) can be held by the flowrate of the bottoms, the vapor boilup, or the feed (if the feed is partially liquid and the stripping section does not contain too many trays). Since the typical hydraulic lag is 3 to 6 seconds per tray, a 20-tray stripping section introduces a deadtime of 1 to 2 minutes in the feed-to-base-level loop. Because of these hydraulic lags, reflux is only very rarely used to control base level. For this loop to work successfully, the column must be relatively short (less than 30 trays) and the holdup in the base must be large (more than 10 minutes).

A fourth degree of freedom is consumed to control column pressure. The valves available are condenser cooling (by far the most commonly used), reboiler heat input, and feed (if the feed is partially vapor). If a flooded condenser is used, the cooling water valve is wide open and an additional valve, typically located between the condenser and the reflux drum, is used to cover or expose heat-transfer area in the condenser.

Finally we are left with two remaining degrees of freedom. So we can control two other variables, for example, two compositions, two temperatures, or one flowrate and one temperature. Typical selections are described in Sect. 6.4. Ideally, since the column is supposed to provide a specified separation between the light key component (LK) and the heavy key component (HK), we would like to control the amount of heavy impurity in the distillate product $x_{D,HK}$ and the amount of light impurity in the bottoms product $x_{B,LK}$. Direct composition measurements are available on some columns, but on-line analyzers are expensive and often unreliable. Therefore, many columns are operated using temperatures to infer compositions. Section 6.5 discusses this in more detail.

It is important to note that during this discussion of degrees of freedom, we have said nothing about the number or types of chemical components involved. If we are separating an ideal binary mixture, we have two degrees of freedom for composition control. If we are separating a multicomponent nonideal mixture, we still have only two degrees of freedom.

Also there are restrictions on what variables we can control. For example, in a multicomponent system we cannot independently specify that two compositions in the same stream are to be controlled. To illustrate this, consider the case where the feed contains a light key component LK, a heavy key component HK, and a heavier-than-heavy key component HHK. We cannot specify the impurity levels of both the light key and the heavier-than-heavy key in the bottoms product. If some of the light key goes out the bottom of the column, essentially all of the heavier-than-heavy key in the feed stream will also go out the bottom no matter what we do. So the composition $x_{B,HHK}$ cannot be controlled.

Another infeasible selection of the two degrees of freedom is to use one of them to control a composition somewhere in the column and to use the other to control the flow of either product stream D or B. Fixing the flowrate of either product stream sets the overall mass and component balances and limits the feasible range of product compositions that satisfy the overall component balances.

This effect is best explained by a simple illustration. Suppose we feed a column with 50 mol/h of A and 50 mol/h of B, and A is the more volatile component. Suppose the distillate contains 49 mol/h of A and 1 mol of B, and the bottoms contains 1 mol/h of A and 49 mol/h of B. Thus the distillate flowrate is $D = 50$ mol/h and the purity of the distillate is $x_{D,A} = 0.98$. Now we attempt to fix the distillate flowrate at 50 mol/h and also hold the distillate composition at 0.98 mole fraction A. Suppose the feed composition changes to 40 mol/h of A and 60 mol/h of B. The distillate will now contain almost all of the A in the feed (40 mol/h), but the rest of it (10 mol/h) must be component B. Therefore the purity of the distillate can never be greater than $x_{D,A} = 40/50 = 0.80$ mole fraction A. The overall component balance makes it impossible to maintain the desired distillate composition of 0.98. We can go to infinite reflux ratio and add an infinite number of trays, and distillate composition will never be better than 0.80.

Any control structure that attempts to fix one of the product flowrates and also attempts to control any composition or temperature in the column will not work if any disturbance occurs in feed flowrate or feed composition.

6.3.2 Fundamental composition-control manipulated variables

In the context of partial control, as we discussed for reactors, we must also identify the dominant variables in the unit and their relationship with the available manipulators.

A useful way to think about the remaining two degrees of freedom that can be utilized for composition control is to classify them as two fundamental manipulated variables: fractionation and feed split.

1. *Fractionation* means the amount of energy and the number of stages used to achieve the separation. For a column with a fixed number of trays, fractionation is reflected by the reflux ratio or the reboiler heat input. Energy (exergy) is needed to provide the thermodynamic work of separation for the components in the feed mixture, as we discussed previously. Fractionation can be set directly and explicitly by using reboiler heat input or reflux ratio to control one composition (or temperature). Or it can be set indirectly and implicitly by, for example, using vapor boilup to hold base level and manipulating bottoms product to control a composition (or temperature).

2. *Feed split* means the fraction of the feed that leaves in one product stream, e.g., the D/F ratio. Feed split can be set directly and explicitly by using either D or B to control one composition. Or it can be set indirectly and implicitly by using reflux or vapor boilup to control one composition and removing D or B to hold reflux drum or base level.

The impact of feed split is much greater than that of fractionation on composition provided that enough energy is available to make the products reasonably pure. The discussion in the previous section about the infeasibility of fixing a product flow and trying to control composition illustrated this point. Another way to look at the effects of feed split and fractionation is illustrated in Fig. 6.7. Changing feed split moves the compositions (and temperatures) up and down in the column, while changing the fractionation sharpens or flattens the composition (and temperature) profile in the column. Decreasing the D/F ratio pushes more light component down the column and causes a large increase in the amount of light-component impurity in the bottoms. The distillate becomes slightly more pure. Decreasing fractionation (reflux ratio) results in smaller changes in compositions from tray to tray (a less sharp composition or temperature profile) and increases the impurity levels of both product streams.

However we choose to look at it, a basic distillation column has two control degrees of freedom. When we turn to more complex column configurations with sidestreams, side strippers, side rectifiers, intermediate reboilers and condensers, and the like, we add additional control degrees of freedom. These more complex systems are discussed in Sec. 6.8.

Decrease D/F

Decrease RR

x_D 0.98 → 0.99

x_B 0.02 → 0.06

x_D 0.98 → 0.975

x_B 0.02 → 0.025

Figure 6.7 Relative effects of feed split and fractionation.

6.3.3 Constraints

One of the challenging aspects of distillation column control is the many limitations imposed on the operation of the column. There are hydraulic constraints, separation constraints, heat-transfer constraints, pressure constraints, and temperature constraints. We recommend the excellent books by Kister (1992 and 1990) on distillation design and operation.

1. *Hydraulic constraints:* A distillation column is a complex fluid flow system. The bottom of the column has the highest pressure in the system so that the vapor can flow from the bottom up through each tray. There is a negative pressure gradient as we move up the column. The pressure drop through each tray depends upon the vapor flowrate and the height of liquid on the tray (weir height plus height over the weir). The liquid must flow down the column against a positive pressure gradient. This is achieved by using the density difference between the liquid and vapor phases to build up enough liquid in the downcomer to overcome the static pressure difference. If the vapor flowrate is too high, the column will flood (liquid cannot flow down the column because of the large pressure drop through the trays). If the liquid flowrate is too high, the height over the weir becomes too high and the column floods because the increase in pressure drop fills the downcomer. Problems also are encountered if vapor and liquid rates become too low. If there is not enough vapor flowing through the holes on the trays to support the liquid, liquid

will weep or dump down through the holes. If there is not enough liquid flowing across the tray, the liquid may be poorly distributed and dry spots may occur. In both cases fractionation will be poor. So distillation column control systems usually have maximum and minimum constraints on the heat input and on the reflux flow.

2. *Separation constraints:* The separation in a column can be expressed as the impurity levels of the key components in the two products: $x_{B,LK}$ in the bottoms and $x_{D,HK}$ in the distillate. Separation is limited by the minimum reflux ratio and the minimum number of trays. We must always have more trays than the minimum and a higher reflux ratio than the minimum. If the number of trays in the column is not large enough for the desired separation, no amount of reflux will be able to attain it and no control system will work. In extractive distillation columns, there is also a maximum reflux ratio limitation, above which the overhead stream becomes less pure as the reflux increases.

3. *Heat transfer constraints:* Heat must be transferred into the liquid in the reboiler to boil off the vapor needed to provide the vapor-liquid contacting in the column. If the base temperature becomes too high and approaches the temperature of the heating source, the heat transfer rate will decrease and vapor boilup will drop. The same result occurs if the reboiler fouls and the heat transfer coefficient drops. In the condenser, heat must be transferred from the hot vapor into the coolant stream to remove the heat of condensation. If the column is operating at its maximum pressure, capacity may be limited by condenser heat removal.

4. *Pressure / temperature constraints:* Pressures and temperatures cannot approach critical conditions because flow hydraulics depend upon the density difference between the liquid and vapor phases. Thermally sensitive components require that temperatures be held below some maximum level. Since the highest temperature occurs in the reboiler, base temperatures may have to be monitored and limited.

All of these constraints mean that the control system must keep the column inside a feasible operating window that may be quite narrow or rather wide.

6.4 Typical Control Schemes

A number of alternative structures are used to control distillation columns. In this section we present some of the most commonly employed strategies and discuss when they are appropriate. The standard terminology is to label a control structure with the two manipulated variables that are employed to control compositions. For example, the "R-V" structure refers to a control system in which reflux and vapor boilup are used to control two compositions (or temperatures). The "D-V" structure means distillate and vapor boilup are used.

The simultaneous control of two compositions or temperatures is called *dual composition* control. This is ideally what we would like to do in a column because it provides the required separation with the minimum energy consumption. However, many distillation columns operate with only one composition controlled, not two. We call this *single-end composition* control.

This is due to a variety of reasons. Dual requires two controllers that interact, making them more difficult to tune. Often the difference in energy consumption between dual and single-end composition control is quite small and is not worth the additional complexity. Frequently direct measurement of composition is difficult, expensive, and unreliable, so temperatures must be used. The column temperature profile may permit only one temperature to be used for control because of the nonuniqueness of temperature in a multicomponent system, resulting in a lack of sensitivity to changes in column conditions. Perhaps the most important reason that most columns operate with single-end control is that just one tray temperature is a dominant variable for column behavior. The dominant temperature usually occurs either in the stripping or rectifying section where there is a significant break in the temperature profile. Controlling this single dominant variable generally provides partial control of both product compositions in the column. Therefore we often use the R-V structure, for example, with reflux flow controlled and reboiler heat input used to control an appropriate tray temperature. More discussion of this subject is given in Sec. 6.6.

Figure 6.8 shows a number of control configurations for simple two-product distillation columns.

1. R-V: Reflux flow controls distillate composition. Heat input controls bottoms composition. By default, the inventory controls use distillate flowrate to hold reflux drum level and bottoms flowrate to control base level. This control structure (in its single-end control version) is probably the most widely used. The liquid and vapor flowrates in the column are what really affect product compositions, so direct manipulation of these variables makes sense. One of the strengths of this system is that it usually handles feed composition changes quite well. It also permits the two products to be sent to downstream processes on proportional-only level control so that plantwide flow smoothing can be achieved.

2. D-V: If the column is operating with a high reflux ratio (RR > 4),

the D-V structure should be used because the distillate flowrate is too small to control reflux drum level. Small changes in vapor to the condenser would require large changes in the distillate flowrate if it is controlling level. When the D-V structure is used, the tuning of the reflux drum level controller should be tight so that the changes in distillate flowrate result in immediate changes in reflux flowrate. If the dynamics of the level loop are slow, they slow down the composition loop. One way to achieve this quick response is to ratio the reflux to the distillate and use the level controller to change the ratio.

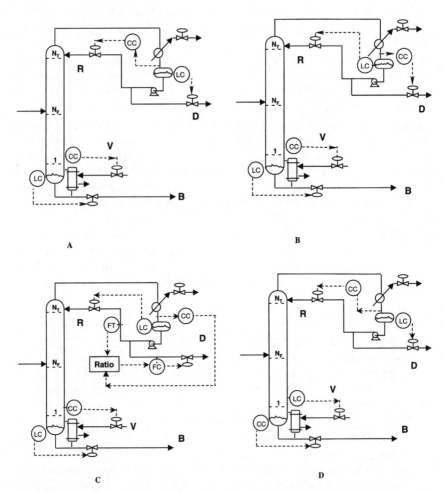

Figure 6.8 Common control structures for distillation columns. (a) Reflux-boilup; (b) distillate-boilup; (c) reflux ratio-boilup; (d) reflux-bottoms; (e) reflux ratio–boilup ratio.

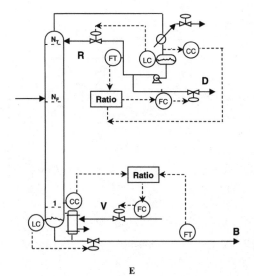

Figure 6.8 *(Continued)*

3. RR-V: Reflux ratio is used to control distillate composition and heat input controls bottoms composition.

4. R-B: When the boilup ratio is high (V/B), bottoms flow should be used to control bottoms composition and heat input should control base level. However, in some columns potential inverse response may create problems in controlling base level with boilup.

5. RR-BR: Reflux ratio controls distillate composition and boilup ratio controls bottoms composition.

Figure 6.9 shows typical control structures for two special types of columns. Figure 6.9*a* is for a column whose feed contains a small amount of a component that is much more volatile than the main component. The distillate product is a small fraction of the feed stream. It is removed from the reflux drum as a vapor to hold column pressure. Reflux flow is fixed, and reflux drum level is controlled by manipulating condenser coolant. In the petroleum industry, this type of column is called a *stabilizer*. The first column in the HDA process is this type.

Figure 6.9*b* shows a column that is separating a mixture with a low relative volatility, so the column has a large number of trays and operates with a high reflux ratio. This type of column is called a *superfractionator*. Because of the high reflux ratio, reflux should be used to control reflux drum level. For the same reason, vapor boilup should be used to control base level. Therefore the two manipulators left to control composition are distillate and bottoms flowrates. Obviously these two

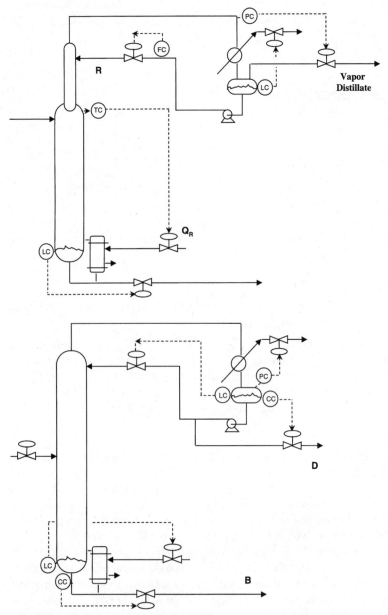

Figure 6.9 Common types of columns and controls. (*a*) Stabilizer (small distillate flow); (*b*) superfractionator with distillate-bottoms control structure.

flows cannot be set independently for a given feed under steady-state conditions. Dynamically, however, they can be adjusted independently. This D-B control structure works well on this type of column. It should be noted that it is quite "fragile" because if either of the two composition loops is put on manual, the other cannot work. Override controls must be used to recognize that this situation has occurred and switch the control structure. For example, if the bottoms composition analyzer fails, the control structure should be switched so that overhead composition is controlled by distillate flow, base level is controlled by bottoms flow, and reboiler heat input is constant.

6.5   Inferential Composition Control

Many industrial columns use temperatures for composition control because direct composition analyzers can be expensive and unreliable. Although temperature is uniquely related to composition only in a binary system (at known pressure), it is still often possible to use the temperatures on various trays up and down the column to maintain approximate composition control, even in multicomponent systems. Probably 75 percent of all distillation columns use temperature control of some tray to hold the composition profile in the column. This prevents the light-key (LK) impurities from dropping out the bottom and the heavy-key (HK) impurities from going overhead.

How do we select the "best" tray location for this temperature control? The next section outlines the normal procedure and illustrates its application with a specific example.

6.5.1   Criteria for selection of best temperature control tray

Since we have two control degrees of freedom, our objectives in distillation are to control the amount of LK impurity in the bottoms product $(x_{B,LK})$ and the amount of HK impurity in the distillate $(x_{D,HK})$. Controlling these compositions directly requires that we have composition analyzers to measure them. Instead of doing this, it is often possible to achieve fairly good product quality control by controlling the temperature on some tray in the column and keeping one manipulated variable constant. Quite often the best variable to fix is the reflux flowrate, but other possibilities include holding heat input or reflux ratio constant.

The typical procedure for selecting which tray to control is to look at the steady-state temperature profile in the column at the base-case conditions, as illustrated in Fig. 6.10. We look for a location in the column where there are large temperature changes from tray to tray. In Fig. 6.10, the slope of the temperature profile is the steepest in the

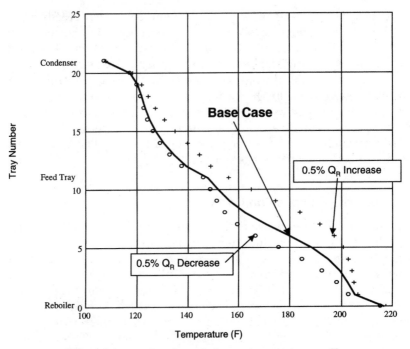

Figure 6.10 Effect of changes in column boilup on temperature profile.

stripping section of the column. Then we select the tray where the temperature profile is changing the most from tray to tray. This is somewhere around trays 6, 7, or 8 in this example. This "steepest slope" criterion is often used to pick the location of the control tray.

But why is this a good way to select the "best" control tray? It works because we want to find a tray where temperature is significantly affected by the compositions of the LK and HK components. Since temperature is also affected by other variables (pressure and other components), it is important that the effects of these variables are small compared with the effects of the compositions of the key components.

Other criteria are also commonly used to guide the selection of the control tray. One is to find the control tray that is most sensitive to changes in the manipulated variable. Figure 6.10 illustrates this method. A steady-state rating program is used to make small open-loop changes in the manipulated variable (heat input in this example). We look at the resulting changes in tray temperatures and select the tray that shows the largest change. In the numerical case shown in Fig. 6.10, tray 6 appears to be the location where changes in heat input (Q_R) produce the largest changes in tray temperature. This procedure gives the largest steady-state gain between the controlled variable and manipulated variable.

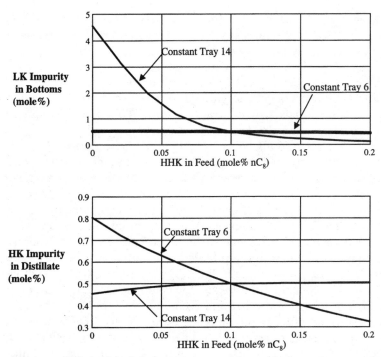

Figure 6.11 Effect of feed composition on product purities with constant tray temperatures.

A third criterion is to find the control tray where the steady-state responses are similar for both positive and negative changes in the manipulated variable. This procedure attempts to avoid problems with nonlinearity. Figure 6.10 shows that tray 6 does a good job in satisfying this criterion.

A fourth criterion is to find the tray location that achieves the fundamental objective of controlling product compositions. The steady-state rating program is run for different values of feed composition, while holding a specified tray temperature constant. A plot such as Fig. 6.11 gives results for our specific example with two different control tray locations (trays 6 and 14) as feed composition is varied. Using a control tray near the bottom of the column produces good control of bottoms composition but not of top composition. The reverse is true when tray 14 near the top of the column is used. So the selection of which tray is best depends upon which stream composition is more important from a product-quality standpoint. Note that the variability in the bottoms product with constant tray 14 temperature is very much larger than the variability of the distillate product with constant tray 6 temperature. This indicates that tray 6 is the better choice in most cases, which is consistent with the previous criteria.

A final method for selecting a temperature control tray location is to use singular value decomposition (SVD) techniques. This approach was first presented by Downs and Moore and is summarized on p. 458 in Luyben and Luyben (1997). A steady-state rating program is used to obtain the gains between the two manipulated variables and the temperatures on all trays. The gain matrix is decomposed by using SVD to find the "most sensitive" tray locations. This method requires more computation than the others.

The criteria discussed above are usually effective because most columns must deal with feeds that are not simple binary mixtures and with column pressures that are not constant. If the system were binary and if pressure were constant, temperature would be directly related to composition. However, neither of these conditions is typically found in operating distillation columns. Even if the pressure at the top of the column is held constant, pressure variations occur down in the column because changes in vapor and liquid flowrates alter the pressure drop through the trays. Most columns have lighter-than-light key components (LLK) and heavier-than-heavy key components (HHK) in addition to the key components (LK and HK), and their presence distorts the temperature profile. Light components in the feed cause a slight drop in the temperatures up through the entire rectifying section, and, since they build up near the top of the column, they can cause a significant drop in temperature near the top of the column. The magnitude of this decrease in temperature depends upon the amount of the LLK in the feed stream. The reverse effect is felt in the stripping section and in the base of the column with HHK components in the feed as temperatures are increased.

Figure 6.12 illustrates this effect. As more LLK component comes in with the feed stream, more depression occurs in the rectifying section temperature profile. If we control a tray temperature near the top and the amount of LLK in the feed increases, the temperature on this tray will start to go down. We will increase heat input to drive it back to its setpoint value, and this will push more HK component out the top. Therefore holding a constant temperature on a tray near the top of the column would result in significant variations in the amount of heavy key component in the distillate product. All of the LLK components must go out the top of the column, and there is nothing we can do about it once these components enter the column. Action must be taken in the upstream column to keep LLK components out of this column. Similar effects occur in the stripping section and near the base when variations occur in the amount of HHK components in the feed. Now temperatures rise as more heavy components enter the column, and we drop more LK component into the bottoms product if we hold a constant temperature on a tray near the base of the column.

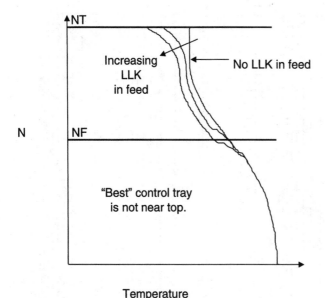

Temperature

Figure 6.12 Effect of lighter-than-light key components on temperature profiles.

Pressure variations can be compensated for by measuring both temperature and pressure at a tray location and computing a pseudocomposition signal. This computed composition signal (pressure-compensated temperature) can then be controlled:

$$x = f_{(T,P)} \tag{6.16}$$

$$\Delta x = \frac{\partial x}{\partial T}\Delta T + \frac{\partial x}{\partial P}\Delta P \tag{6.17}$$

$$= K_1 \Delta T - K_2 \Delta P \tag{6.18}$$

6.5.2 Numerical example

The results shown in Fig. 6.10 were generated for a 20-tray, 2-ft-diameter column, with a feed on tray 11 of 100 lb-mol/h at 90°F. The feed composition is 5 mol % propane, 40 mol % normal butane, 45 mol % normal pentane, and 10 mol % normal octane. Product purity specifications are 0.5 mol % butane in the bottoms and 0.5 mol % pentane in the distillate. The operating pressure of the column is 73 psia, and the reflux ratio is 1.76. The temperature on tray 6 is 179.9°F and on tray 14 is 130.6°F. Figure 6.13 gives the composition profiles in the column. Note the buildup of the HHK component near the bottom of the column

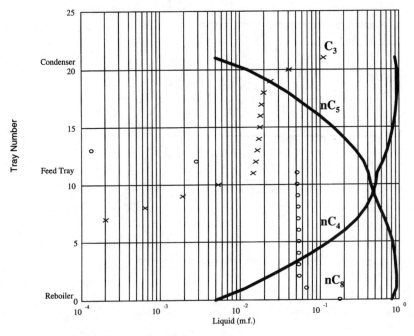

Figure 6.13 Composition profiles for base case.

and the buildup of the LLK component near the top. The effect of these non-key components can be seen in the "tails" on the temperature profile given in Fig. 6.10.

A dynamic simulation of this column using HYSYS was used to explore the dynamics of the process for the two cases where different tray temperatures are controlled. Either tray 6 or tray 14 temperature is controlled by manipulating reboiler heat input. Reflux flowrate is held constant. Disturbances are step changes at time equals 5 minutes in feed flowrate (25 percent increase) or feed composition. The feed composition disturbance is a drop in the HHK component in the feed (normal octane changed from 10 mol % to 0 mol % while normal pentane changed from 45 mol % to 55 mol %).

Figures 6.14 and 6.15 give dynamic responses of the tray temperatures, reboiler heat input, and bottoms product impurity. The temperature loops were tuned using the TL (Tyreus-Luyben) tuning rules after the ultimate gain and ultimate frequency had been determined using a relay-feedback test. Two 0.5-minute first-order lags are used in the temperature loop. Temperature transmitter spans are 100°F. The ultimate gain and period for the tray 6 temperature loop are 4.2 and 2.7 minutes, and for the tray 14 loop are 12.7 and 2.5 minutes. These results reflect the fact that the process gain is higher when tray 6 is

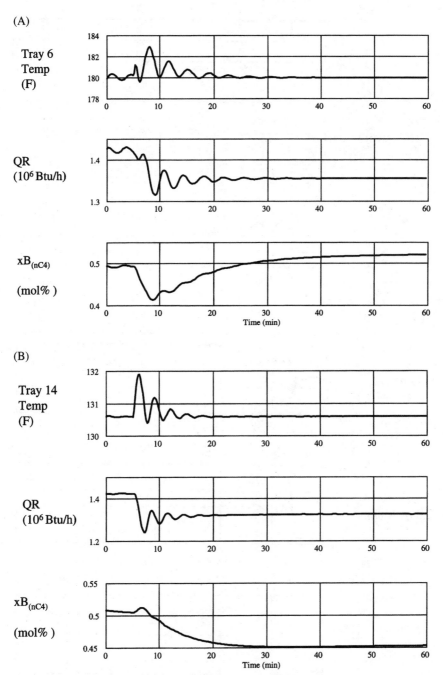

Figure 6.14 Dynamic response to HHK feed composition disturbance. (*a*) With control tray 6; (*b*) with control tray 14.

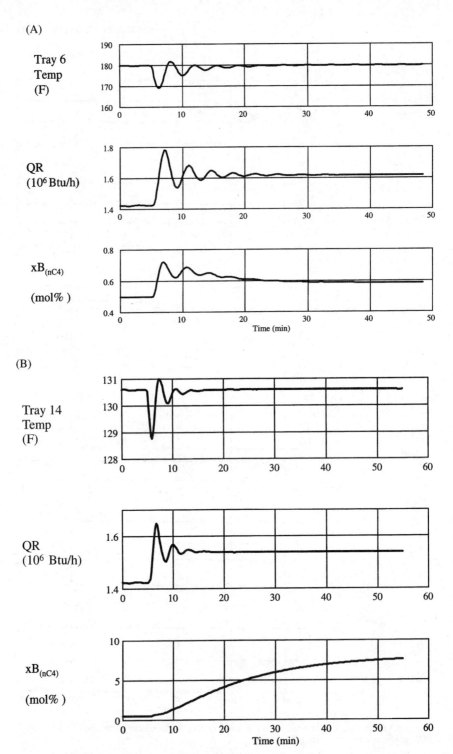

(A)

Tray 6
Temp
(F)

QR
(10^6 Btu/h)

$xB_{(nC4)}$

(mol%)

Time (min)

(B)

Tray 14
Temp
(F)

QR
(10^6 Btu/h)

$xB_{(nC4)}$

(mol%)

Time (min)

Figure 6.15 Dynamic response to 25 percent increase in feed flow. (a) With control tray 6; (b) with control tray 14.

212

controlled, requiring a lower controller gain. The dynamics of the loops are essentially the same because vapor rates affect temperatures on all trays very quickly.

As shown in Fig. 6.15, the 25 percent increase in feed flowrate, with the reflux flowrate constant, results in a slight increase in the impurity in the bottoms product when tray 6 is controlled. But when tray 14 is controlled, there is a very large change in bottoms composition.

These dynamic simulations confirm that the control tray selection criteria discussed earlier yield reasonably effective control systems.

6.5.3 Flat temperature profiles

The use of a single tray temperature is only viable if there is a section of the column where sufficient changes in temperature occur from tray to tray. In difficult (low relative volatility) separations, there is very little change in temperature from tray to tray. The effects of pressure and feed composition variations can swamp the effect of key component compositions on this flat temperature profile. If the temperature change between the top and the bottom of the column is less than about 20 to 30°F, a single tray temperature may prove to be ineffective for control.

In attempts to overcome the problem of flat temperature profiles, differential temperature has been used in some systems. Looking at the difference in temperature between two trays helps to reduce the influence of pressure on the temperature signals. However, it does not handle changes in pressure drop.

In addition we must be aware that the relationship between ΔT and product composition is nonmonotonic, and we must know on which side of the peak we are operating. Figure 6.16 illustrates the situation. When the temperature profile is above the section in which the temperature difference is being measured, that section is filled with heavy component, and the temperature difference is small. When the temperature profile is below the section in which the temperature difference is being measured, that section is filled with light component, and the temperature difference is small. In between these two extremes there is a significant temperature difference.

This nonmonotonic behavior can lead to feedback control instability. On one side of the hump the controller should be direct acting. On the other it should be reverse acting. So we must be careful when applying differential temperature control structures.

6.5.4 Sharp temperature profiles

At the other extreme, temperature profiles are very sharp (large temperature differences between trays) when the separation is easy (high relative volatility). This situation also can cause control problems. A

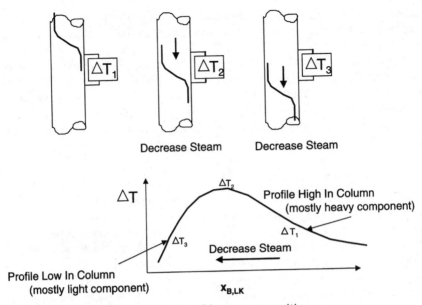

Figure 6.16 Relationship between ΔT and bottoms composition.

very small change in the manipulated variable gives a very large change in temperature. This high process gain requires that the controller gain be small, and this leads to poor load rejection. Saturation of the temperature signal is also a problem. Suppose the temperature profile moves up past the control tray and continues to move up the column. We do not see any change in the temperature signal. Therefore the error signal that the controller sees does not change, and the only mechanism for driving the profile back down the column is the integral action in the controller. This may take a long time, and load rejection can be very poor, particularly since the controller gain is small.

A fairly easy solution to this problem is to use multiple temperature measurements over a section of the column in which the temperature profile moves and control an "average temperature" (Fig. 6.17).

6.5.5 Soft sensors

We started this section on inferential composition control by using a single tray temperature. Then we used two temperatures to control a temperature difference. Then we used multiple temperatures for average temperature control. The logical extension of this approach is to use whatever measurements are available to estimate product compositions. There are several types of composition estimators. Steady-state

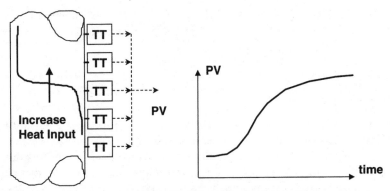

Figure 6.17 Average temperature (profile position) control.

estimators use the steady-state algebraic equations describing the system to predict compositions. Dynamic estimators use dynamic models to predict product compositions. One example of the latter type is the Luenberger Observer. Figure 6.18 illustrates the approach.

A rigorous nonlinear dynamic model of the column is used on-line to predict compositions. The measured flowrates of the manipulated variables (reflux and heat input) are fed into the model. The differential equations describing the system are integrated to predict all compositions and tray temperatures. The predicted tray temperatures are compared with the actual measured tray temperatures, and the differences

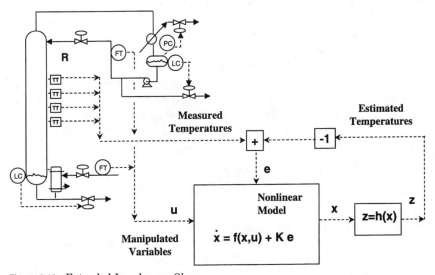

Figure 6.18 Extended Luenberger Observer.

are fed back into the model to drive the predicted temperature to match the measured. The resulting predicted product compositions can be used for product quality control.

6.6 High-Purity Columns

Distillation columns that produce products with parts-per-million (ppm) levels of impurities offer some challenging control problems. These columns exhibit very nonlinear responses to changes in manipulated variables and disturbances. The nonlinear effects take two forms: nonlinear steady-state gains and nonlinear dynamics.

Changing an input variable in one direction can produce small changes in product composition, while changing it in the other direction can produce very large changes in product composition. This asymmetric behavior of the steady-state gains can result in sluggish control in one direction and oscillatory control in the other direction.

Small changes in input variables take a very long time to be felt (large time constants), but larger changes are felt quickly. Time constants predicted by a linear model (which assumes infinitesimal perturbations) can be in the order of days for impurity levels in the 10-ppm range. Because of the low concentrations in the products, gas chromatographs usually must be used for on-line analysis, and these devices have significant deadtimes. Looking at the huge time constants predicted by a linear model, we might expect that the response of the column to disturbances is very slow, and therefore it is not important to be able to detect changes quickly in product compositions. This is not true. It is important to have small chromatographic sampling times so that the rapidly felt large disturbance can be detected quickly and corrective action taken.

Despite their highly nonlinear nature, high-purity columns can be effectively controlled by conventional linear controllers. If the composition control loop can be tuned for tight control (which requires small deadtimes), the column composition profile can be kept in the proper location and products will not go off specification. However, if the controller is loosely tuned and the profile is permitted to move either up or down, very large changes in product purities will occur. The controller must be tuned tight enough to keep the column from "going over the cliff" into the nonlinear region.

Since high-purity products are usually produced only in situations where the separation is relatively easy, most of these columns have fairly large temperature gradients. Therefore it is possible to use temperature/composition cascade control systems. The secondary temperature controller serves as a fast loop to detect disturbances quickly and hold the temperature profile in the column. The primary composition controller provides a trim to the setpoint of the temperature controller to adjust the temperature profile so that product purity is maintained.

Dual composition control is not recommended for high-purity columns. It is better to select one end or the other for control and provide sufficient reflux to handle the worst-case conditions such that the purity of the uncontrolled product is always at or above specification.

6.7 Disturbance Sensitivity Analysis

We can gain some very useful knowledge about the effectiveness of alternative control structures by using a steady-state analysis of the column. The idea is to use a steady-state rating program for the column to calculate how the manipulated variables must change in the face of disturbances for a given control structure. If these steady-state changes are small, the proposed control structure has a good chance of providing effective control of the plant.

For example, let us consider a simple distillation column in which we have specifications on both the distillate and bottoms products ($x_{D,HK}$ and $x_{B,LK}$). We go through the design procedure to establish the number of trays and the reflux ratio required to make the separation for a given feed composition. This gives us a base case from which to start. Then we establish what disturbances will affect the system and over what ranges they will vary. The most common disturbance, and the one that most affects the column, is a change in feed composition. Next we propose a partial control structure. By *partial* we mean we must decide what variables will be held constant. We do not have to decide what manipulated variable is paired with what controlled variable. We must fix as many variables as there are degrees of freedom in the system of equations.

Next we use a steady-state simulator to calculate the values of all the dependent variables for each feed composition while holding $x_{D,HK}$ and $x_{B,LK}$ constant. Finally we look at the steady-state changes in the manipulated variables. If these changes are reasonably small, the proposed control structure may provide good control. We can draw this conclusion from the following argument.

The output signal from a PI controller is the sum of two terms: the error term and the integral of the error term. At the new steady state, the error must be zero. Therefore the integral of the error is directly related to the change in the controller output, which is the change in the manipulated variable.

$$\Delta \, CO = \Delta \, (\text{manipulated variable}) = \frac{K_c}{\tau_I} \int E \, dt \qquad (6.19)$$

The larger the integral of the error is, the poorer the control. The controller must work harder if it has to drive its output to a new steady state that is far from where it started. Thus the steady-state changes in the manipulated variables are directly correlated to control effectiveness.

This steady-state analysis tells us only where we are going. It does not tell us the path we travel to get there. Hence it provides a *necessary but not sufficient* condition test for a proposed control structure. The structure may have dynamic problems that make it a poor performer, even though the steady-state analysis looks good.

Disturbance sensitivity analysis can be used in two ways: (1) to screen proposed control structures and (2) to suggest simplified control structures. As an example of the second function we return to the single-column process discussed above. We want to control both $x_{D,HK}$ and $x_{B,LK}$. We could consider doing this by using a dual composition control system. However, suppose the disturbance sensitivity analysis shows that the changes in the reflux flowrate are quite small when feed composition varies over its expected range. This suggests that a single-end control scheme might provide quite good control and waste very little energy. We could simply fix the reflux flowrate (or ratio it to the feed rate to the column to handle throughput changes) and control a single composition (or temperature) somewhere in the column by manipulating heat input.

6.8   Complex Columns

In this section we present more complex distillation column processes that go beyond the "plain vanilla" variety. Industry uses columns with multiple feeds, sidestreams, combinations of columns, and heat integration to improve the efficiency of the separation process. Very significant reductions in energy consumption are possible with these more complex configurations. However, they also present more challenging control problems. We briefly discuss some common control structures for these systems.

6.8.1   Sidestream Columns

The conventional flowsheet to separate a ternary mixture uses two distillation columns in series. It is sometimes more economical to use a single distillation column with a sidestream. This is particularly true when product purities are moderate to low. Consider the case where the ternary mixture contains components A, B, and C, with decreasing relative volatilities. Figure 6.19 shows two common situations: a liquid sidestream is withdrawn from a tray somewhere above the feed tray, or a vapor sidestream is withdrawn from a tray somewhere below the

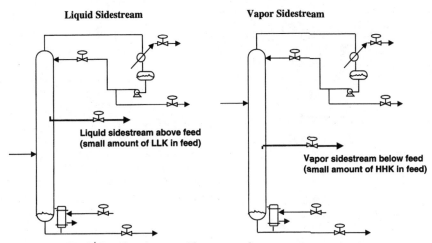

Figure 6.19 Single liquid and vapor sidestream columns.

feed tray. Component A comes out in the distillate product, component B in the sidestream, and component C in the bottoms.

Liquid sidestreams are used when there is less A in the feed than B or C. The vapor passing up through the sidestream drawoff tray contains all the A that is in the feed. The concentration of A in the liquid phase on this tray is smaller than its concentration in the vapor phase since A is the most volatile component. Therefore we withdraw a liquid sidestream. In the case where there is less C in the feed than A or B, we use a vapor sidestream below the feed. The concentration of the sidestream impurity C is less in the vapor phase than in the liquid phase since C is the heaviest component.

Figure 6.20 gives a control scheme for a single sidestream column. The flowrate of the sidestream can be manipulated, so we have an additional control degree of freedom. Three compositions can be controlled: the impurity of B in the distillate $(x_{D,B})$, the purity of B in the sidestream $(x_{S,B})$, and the impurity of B in the bottoms $(x_{B,B})$. Note that we cannot control the two impurity levels (A and C) in the sidestream $x_{S,A}$ and $x_{S,C}$ because there are not enough degrees of freedom.

Sidestream columns are often used in cases with very small amounts of either the lightest component A or the heaviest component C. Here the objective is to purge these small amounts of impurity. The simplest way to do this is to fix the flowrate of the purge stream at a high enough level so that in the worst case (when the maximum amount of the impurity is present in the feed) it will all go out in the purge stream. Some of the desired product is lost in the purge. But if the amount of the component to be purged is very small in the feed, the yield loss is

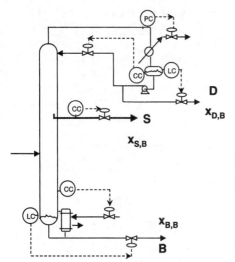

Figure 6.20 Sidestream column control.

small. Figure 6.21 gives control schemes for the two cases: a light purge and a heavy purge. The purge flowrates are fixed (or ratioed to column feed rate). The two main products must be removed to satisfy the component balances around the column. A ratio scheme is shown for the light purge column: sidestream is ratioed to reflux flowrate. This

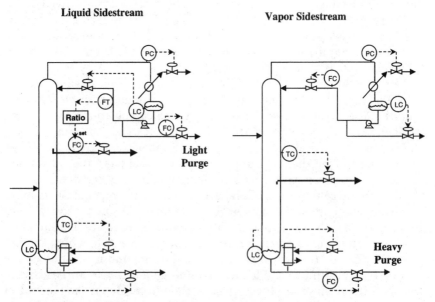

Figure 6.21   Purge column control with liquid or vapor sidestreams.

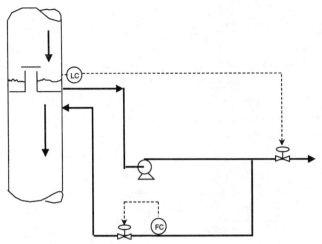

Figure 6.22 Sidestream total trap-out tray.

permits the sidestream flowrate to vary as the amount of component *B* in the feed changes.

Problems are frequently encountered in sidestream columns when the sidestream flowrates represent a large fraction of the total liquid (or vapor) traffic through that section of the column. If the flowrate of a liquid sidestream is more than 50 percent of the total liquid entering the sidestream drawoff tray, the liquid that flows on down the column below the sidestream tray is the difference between two large flows. Thus small changes in either flow produce large changes in the net liquid flowrate down the column, and this flowrate is unmeasured. This can result in drastic changes in the liquid loading of the trays below the sidestream. In severe cases a process change may be required to prevent flooding or weeping. As shown in Figure 6.22, a total liquid trap-out tray can be used. The liquid to the lower section is flow-controlled. The sidestream is withdrawn on level control from the trap-out tray.

Similar problems can occur with vapor sidestreams, but the solution is not as easy because we cannot provide vapor holdup in the system. One approach is to use an *internal vapor* controller. The flowrate of the vapor sidestream and the flowrate of the steam to the reboiler are measured. The net flowrate of vapor up the column above the vapor sidestream drawoff tray is calculated. This flow is then controlled by manipulating the vapor sidestream drawoff rate.

Sidestream columns are also used in combination with other columns. Figure 6.23 gives three common configurations in which sidestream columns are linked to strippers, rectifiers, or prefractionators.

Control of Individual Units

(A)

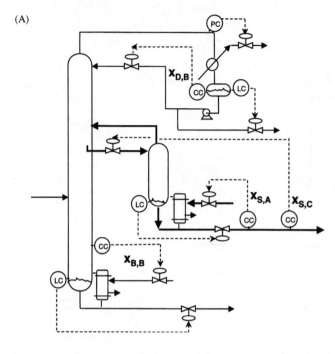

(B)

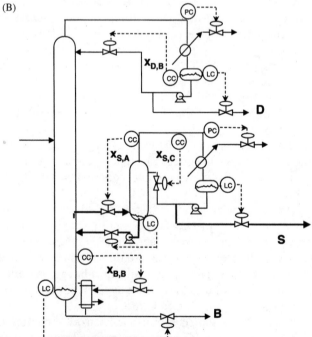

Figure 6.23 Sidestream column with other columns. (*a*) With stripper; (*b*) with rectifier; (*c*) with prefractionator.

(C)

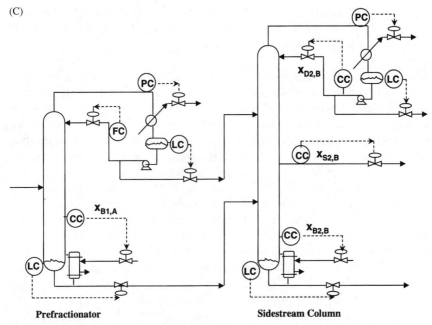

Prefractionator   Sidestream Column

Figure 6.23 *(Continued)*

Sidestream column with stripper.   Higher-purity sidestream products can be obtained if a stripping column is used in conjunction with a side-stream column. The liquid drawoff stream from the main column is fed onto the top tray of a stripper. The stripper has a reboiler, which produces vapor to strip out most of the light component A.

This configuration adds two degrees of freedom to the conventional column (the liquid flowrate to the stripper and the heat input to the stripper reboiler), so two compositions in the sidestream product from the stripper can theoretically be controlled. The control system shown in Fig. 6.23a uses stripper reboiler heat input to control the impurity of A in the sidestream product. The impurity of C in the sidestream product is controlled by manipulating the flowrate of liquid to the stripper. This system presents a highly interacting 4×4 multivariable control problem. Therefore in practice it may be more effective to control only one composition (or temperature) in the stripper and one tempera-ture in the main column, with the flowrate of reflux and liquid to the stripper flow controlled.

Sidestream column with rectifier.   Figure 6.23b shows a process where a vapor sidestream is fed into a rectifying column to remove some of the C impurity in the vapor stream. A 4×4 multivariable control strategy is shown, but practical considerations would probably favor a simpler control structure with the flowrate of vapor to the rectifier on flow control (or ratioed to the vapor boilup).

Sidestream column with prefractionator. Figure 6.23c illustrates a complex configuration in which a prefractionator column is used to perform a preliminary separation of the ternary feed. The idea is to produce a distillate from the first column that contains very little of the heaviest component C. When this distillate is fed into the second column at a location above the sidestream drawoff, there will be only a small amount of C that must flow down past the sidestream tray. This permits the production of high-purity sidestream product. Similarly the prefractionator should let very little of the lightest component A drop out the bottom so that there is little A in the vapor stream flowing past the sidestream tray. This lets us achieve high sidestream purities.

This process adds additional degrees of freedom, which we may want to utilize. The control scheme shown in Fig. 6.23c is simple and practical. Reflux flowrate in the first column is flow-controlled and heat input prevents A from dropping out the bottom of the first column. Heat input to the second column controls the impurity of B in the bottoms product. Sidestream flowrate controls the purity of the sidestream product. Reflux flowrate in the second column controls the impurity of B in the distillate product.

6.8.2 Heat-integrated columns

The tenfold increase in energy prices in the 1970s spurred efforts to reduce energy consumption in chemical and petroleum plants. Heat integration was extensively applied to achieve very significant reductions in energy consumption in distillation columns. There are a host of alternative configurations that have been built in industry. We discuss below several of the most widely used process structures and their control schemes.

Feed split (binary). Figure 6.24a shows a heat-integrated distillation configuration in which a feed stream is split into two streams and fed into two columns, one operating at high pressure and one operating at low pressure. The basic idea is the same as a multieffect evaporator: the energy transferred into the high-pressure column (operating at a high temperature) is reused in the low-pressure column (operating with a base temperature that is lower than the bubble-point temperature of the distillate in the high-pressure column). The required temperature difference in the heat exchanger coupling the two columns (the reboiler of the low-pressure column and the condenser of the high-pressure

(A)

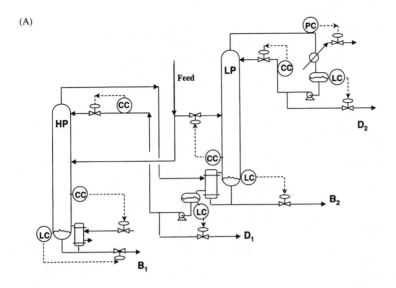

(B)

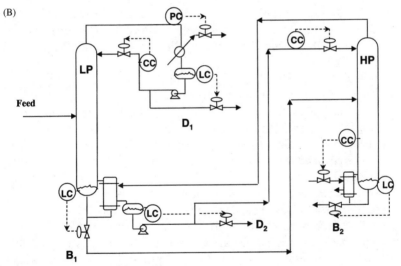

Figure 6.24 Heat-integrated columns. (*a*) Feed split (binary); (*b*) light-split reverse (binary); (*c*) prefractionator reverse (ternary).

column) is typically 20°F in high-temperature applications, but in cryogenic applications may be much smaller because of the high cost of compression.

This structure is normally used only to separate a binary mixture containing no LLK or HHK components. The presence of these components would lower or raise the temperatures at the two ends of the

(C)

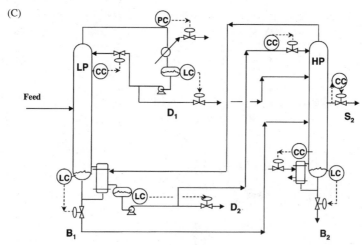

Feed

D_1

S_2

D_2

B_1

B_2

Figure 6.24 *(Continued)*

columns. This would require the use of larger pressure differentials between the columns to achieve the necessary temperature differential driving force for heat transfer and would make the economics less favorable.

The feed split is approximately 50/50, but slightly more feed goes to the low-pressure column if the separation is easier at lower pressure. The system as pictured runs "neat," i.e., all the heat available from condensing the vapor from the high-pressure column is used to reboil the low-pressure column. In some systems auxiliary reboilers and/or condensers are used to balance the heat loads both at steady state and dynamically.

The system represents a 4×4 interacting control problem since there are four product compositions to be controlled at each end of both columns. Reflux flowrates control the distillate purities in each column. Bottoms purity in the high-pressure column is controlled by manipulating the heat input to the reboiler. Bottoms purity in the low-pressure column is controlled by manipulating the fraction of the feed that is fed into the low pressure column.

Light-split reverse. Figure 6.24*b* shows an alternative configuration where about half of the lighter component is removed in the first column and the other half in the second column. The bottoms from the first column is a mixture of light and heavy components and is fed into the second column. The first column can be run at high pressure and the second at low pressure (direct heat integration). In the flowsheet shown, the pressure in the first column is low and in the second high, so the heat integration is the direction opposite the process flows (reverse heat integration).

Now there are only three product streams, so the control problem is 3×3. In the control system shown, reflux flowrates control the distillate purities in each column. Bottoms purity in the high-pressure column is controlled by manipulating the heat input to the reboiler.

Prefractionator reverse (ternary).   Figure 6.24c shows a third alternative flowsheet that combines heat integration with a complex configuration. This system can be used to separate a ternary mixture. In the system shown, the sidestream column is run at high pressure and the prefractionator at low pressure.

Reflux flowrate in the prefractionator is manipulated to keep the temperature profile where it should be, so that very little of the heaviest component goes out the top and very little of the lightest component goes out the bottom. Product purities in the sidestream column are controlled by reflux flowrate, sidestream flowrate, and heat input.

To illustrate the very large energy savings that are possible with this complex/heat-integrated system, consider the separation of a benzene, toluene, and xylene mixture. A conventional two-column light-out-first separation flowsheet with no heat integration uses *twice* the energy that the prefractionator-reverse flowsheet uses.

6.8.3 Extractive distillation

Addition of a third component to act as a heavy entrainer is frequently done to alter the vapor-liquid equilibrium between the two primary components to be separated. Extractive distillation is used when the separation is very difficult (relative volatilities approaching unity) or when azeotropes are present. Suppose we have a binary mixture of components A and B. A high-boiling extracting agent or solvent S is fed into the first column near the top (see Fig. 6.25). One of the components B is preferentially attracted by the solvent and goes out the bottom with the solvent. The mixture of B and S is fed into a second column that separates them. Distillate from the second column is mostly B. Bottoms from the second column is mostly S and is recycled back to the first column. Reflux on the first column is used to keep S from going out the top in the distillate product, which is mostly A.

The control system shown holds the temperature profile in each column by manipulating heat inputs. Enough reflux is used on both columns to keep the product purities above specification. The solvent flowrate is ratioed to the fresh feed flowrate.

Note that the level in the base of the solvent recovery column is *not*

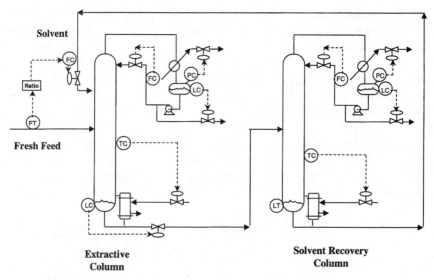

Figure 6.25 Extractive distillation column.

controlled. A small stream of fresh solvent would be added to make up for any losses of solvent in the two product streams over a long period of time, but the base of the column provides the surge capacity for solvent in the system.

It should also be noted that many extractive distillation systems exhibit a *maximum* reflux ratio as well as the conventional minimum reflux ratio. For a given solvent-to-feed ratio, if too much reflux is returned to the column, the solvent is diluted and the separation becomes poorer since not enough solvent is available to soak up component *B*.

6.8.4 Heterogeneous azeotropic distillation

Our final example of a complex column is an azeotropic system in which we add a light entrainer to facilitate the separation of two components. The classical example of this type of system is the use of benzene or cyclohexane to break the ethanol-water azeotrope. As shown in Fig. 6.26, the vapor from the top of the column is condensed and fed into a decanter in which the two liquid phases separate. The aqueous phase is removed as product. The organic phase (the light entrainer) is refluxed back to the column. Some of the organic may also be added to the feed stream to alter the composition profiles in the column (if more entrainer is needed lower in the column). Note that the organic level in the decanter is *not* controlled. A small stream of fresh entrainer would be added to make up for any losses of entrainer over a long period

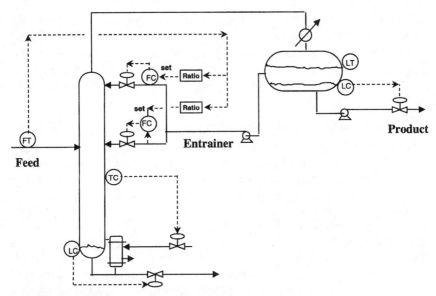

Figure 6.26   Heterogeneous azeotropic distillation column.

of time, but the decanter provides the surge capacity for entrainer in the system.

6.9   Plantwide Control Issues for Distillation Columns

We briefly discussed in Chap. 2 (Sec. 2.5) one important aspect of distillation column control when one of the products leaving the column is a recycle stream. Should the composition of this recycle stream be controlled? Probably not from the perspective of the isolated column because the control loop holding the composition of the other stream leaving the column, which is a product from the plant, could then be more tightly tuned. However, the plantwide control perspective may show that the performance of the reactor can be improved by holding the purity of the recycle stream more constant, and this could result in smaller disturbances to the column. The overall effect may be better product quality control even though the product-quality loop is less tightly tuned.

There are many other aspects of distillation column control in a plantwide context. One of the earliest studies of these issues is the work of Downs (1992). According to Downs, "the control strategy for each unit operation must be developed within the framework of the overall component inventory control structure." He presents several realistically complex reaction/separation examples. This work is the basis of the "Downs drill" discussed in Chap. 3.

In addition to the question of composition control of recycle streams, there are many other examples of plantwide issues affecting the choice of the "best" control structure for a distillation column. Several of these are discussed below.

6.9.1   Reflux drum and base level control

When considering a column in isolation, we typically select heat input to control temperature on an appropriate column tray to achieve good, tight composition control. Then we fix the reflux flowrate or ratio it to the feed flowrate. Finally, we control the level in the reflux drum by manipulating distillate flowrate and control the level in the column base by manipulating bottoms flowrate. This structure works well for the column and also fits nicely into a plantwide context because the proportional-only level controller setting both products provide gradual smooth flowrate changes to downstream sections of the plant. This type of control scheme would be used for a column with a liquid distillate product and with a low to moderate reflux ratio (less than 4).

However, for columns with higher reflux ratios, Richardson's rule dictates that we control reflux drum level with reflux, not distillate. The flowrate of distillate would be used (1) to control distillate product composition or (2) to control a constant reflux ratio. If the former strategy is used (distillate controls composition or temperature), there may be significant fluctuations in the distillate flowrate as the tightly tuned composition controller attempts to achieve good quality control. With this structure the reflux drum level loop must be tightly tuned (PI control) so that we do not introduce an additional lag in the composition loop; i.e., changes in distillate flowrate will result in immediate changes in reflux flowrate.

If the distillate is fed to a downstream unit, the variability in flowrate will be a disturbance. So what can we do? We can make use of feedforward control to anticipate the required changes in reflux and distillate flowrates (ratio distillate to feed with the ratio reset by the composition contoller).

In the second case (manipulate distillate to control reflux ratio), the variability of the distillate flow would be greatly reduced. The reflux drum level controller, manipulating reflux flowrate, is made P-only to get slow changes in reflux flowrate, and this gives slow changes in distillate flowrate in the reflux-ratio control structure.

So from a plantwide control perspective, setting distillate flowrate to control reflux ratio is a better strategy than using distillate to control composition. Of course similar arguments can be made about bottoms flowrate in the case of a column with a high boilup ratio.

6.9.2 Pressure control with vapor distillate product

When a partial condenser is used and the distillate is removed from the column as a vapor, common practice is to use this vapor stream to control column pressure. The reflux drum level is usually controlled by manipulating condenser cooling, and reflux flowrate is fixed or ratioed to feed. Heat input is used to control a tray temperature (Fig. 6.9a).

The swings in the vapor flowrate are typically quite significant to achieve tight pressure control. In an isolated column environment or when this vapor stream simply flows into a large vapor surge vessel or into a large pipeline (header), the variations in the flow cause no problem. But, if this vapor stream is the feed to a downstream unit, the flow variability can significantly disturb this unit and can result in poor plantwide control performance.

So what can we do in this case? If column operation requires that we stick to this control structure, feedforward control will help to reduce the swings in distillate flowrate. However, better plantwide performance can be achieved if we can switch the control structure to one in which the vapor distillate is not used to control pressure. One possible alternative is shown in Fig. 6.27. Condenser cooling is used to control pressure, reflux flowrate controls reflux drum level (with P-only control), and the flowrate of the vapor distillate is ratioed to reflux flowrate. With this structure we allow the disturbances that the column energy

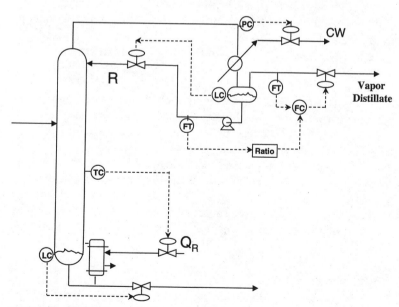

Figure 6.27 Alternative control structure for vapor product.

balance sees to be absorbed by the plant utility system (cooling water flowrate to the condenser). And we use the capacitance of the reflux drum to provide gradual changes in both reflux flowrate and vapor distillate flowrate (through the ratio controller).

6.9.3 On-demand product

Suppose our plantwide control system requires "on-demand" products. As discussed in Chap. 3, the fourth step in the plantwide control design procedure is to establish where production rate is set. If the flowrate of one of the product streams leaving the column is fixed by a downstream unit or customer, the column control structure must be set up in an appropriate way.

Figure 2.2 contrasts a typical column with a variable feed rate as set by an upstream unit to an on-demand column with bottoms flowrate set by a downstream operation. Further examples are given in later chapters when specific plants are considered (Chap. 8 for the Eastman process and Chap. 11 for the vinyl acetate process).

6.9.4 Fresh feed streams for level control

In an isolated column environment, we control reflux drum level and base level with two of the manipulated variables available on the column itself. In a plantwide environment, these levels can also be controlled by the flowrates of fresh feed streams being introduced into the plant. This is done when these levels reflect the inventories of the component within the plant. The ternary process illustrated the scheme in Chap. 2 (Fig. 2.13b). The fresh feed makeup streams F_{0A} and F_{0B} are used to control the levels in the second column reflux drum and in the first column base.

In looking at Fig. 2.13b, it may appear at first glance that the use of these streams to control these levels would yield very poor level control because the flows would have to work their way through the reactor and through the column before the levels see the effects of a change in the fresh feed flowrates. This is *not* the case! There is an immediate effect of changing fresh feed on level. For example, consider the reflux drum on the second column. If the level goes down, the fresh feed F_{0A} is increased. Since the total flowrate of the fresh feed and the distillate ($F_{0A} + D_2$) is flow-controlled, an increase in F_{0A} gives an immediate decrease in D_2. And this affects the level in the reflux drum immediately.

6.9.5 Energy integration

If a column is operating in isolation and using a plant utility for heat input (steam at an appropriate pressure level or fuel if a fired reboiler is used), the manipulation of heat input can be achieved without affecting anything else in the plant. However, if some or all of the heat to a column comes from a hot process stream that is being cooled, the column control structure must be set up so that changes in this heating source do not upset the column. This is an important part of the energy management strategy of the entire plant.

This subject is discussed in detail in Chap. 5, along with the use of auxiliary reboilers and condensers as well as the idea of using "total heat input" controllers.

In many plants high-temperature heat is available from exothermic reactors. This heat can be used in the reboiler of a distillation column operating at lower temperature levels. From a steady-state energy conservation standpoint, it is desirable to thermally link the reactor with the column directly, i.e., transfer the heat from the process fluid in the reactor directly into the process material in the base of the column. This requires one heat transfer area and one temperature difference. Alternatively, steam could be generated in a separate heat exchanger in the reactor, and this steam could then be used in another heat exchanger (the reboiler) in the column. This second alternative requires higher capital investment (more heat transfer area) because the heat must be transferred twice. However, it makes the control of both the reactor and the column much easier and eliminates thermal interaction because the plant steam system can be used to remove or supply steam as needed. So this points out another example of a conflict between dynamic plantwide controllability and steady-state economics.

6.10   Conclusion

In this chapter we have presented some fundamental concepts of distillation control. Distillation columns are without question the most widely used unit operation for separation in the chemical industry. Most final products are produced from one end or the other of a distillation column, so tight control of product quality requires an effective control system for the column. However, the column is usually an integral part of an entire plant, so its control scheme must also be consistent with the plantwide control structure.

Most distillation columns have two control degrees of freedom, once pressure and feed conditions are set. The typical control structure holds the composition profile in the column by controlling a tray temperature somewhere in the column. The other degree of freedom is then normally consumed by fixing some other variable such as the flowrate of reflux, the reflux ratio, or the heat input.

More complex columns (with sidestreams, multiple feeds, intermediate reboiler or condenser, etc.) have more degrees of freedom and require more complex control schemes. Whatever the degree of complexity, it is vital to keep a plantwide perspective when developing control systems for distillation columns.

6.11 References

Agreda, V. H., Partin, L. R., and Heise, W. H. "High-Purity Methyl Acetate via Reactive Distillation," *Chem. Eng. Prog.*, 40–46, February (1990).
Buckley, P. S., Luyben, W. L., and Shunta, J.P. *Design of Distillation Column Control Systems*, Research Triangle Park, N.C.: Instrument Society of America (1985).
Doherty, M. F., and Buzad, G. "Reactive Distillation by Design," *I Chem E* Vol. 70, Part A, 448–458 (1992).
Doherty, M. F., and Malone, M. F. *Conceptual Design of Distillation Systems*, New York: McGraw-Hill (1998).
Downs, J. J. "Distillation Control in a Plantwide Control Environment," Chap. 20 in *Practical Distillation Control*, W. L. Luyben (ed.), New York: Van Nostrand Reinhold (1992).
Kister, H. Z. *Distillation—Design*, New York: McGraw-Hill (1992).
Kister, H. Z. *Distillation—Operation*, New York: McGraw-Hill (1990).
Luyben, W. L. (ed.) *Practical Distillation Control*, New York: Van Nostrand Reinhold (1992).
Luyben, W. L., and Luyben, M. L. *Essentials of Process Control*, New York: McGraw-Hill (1997).
Ognisty, T. P. "Analyze Distillation Columns With Thermodynamics," *Chem. Eng. Prog.*, 40–46, February (1995).
Rademaker, O. J., Rijnsdorp, J. E., and Maarleveld, A. *Dynamics and Control of Continuous Distillation Units*, New York: Elsevier (1975).
Shinskey, F. G. *Distillation Control*, New York: McGraw-Hill (1977).
Sneesby, M. G., Tade, M. O., Datta, R., and Smith, T. N. "ETBE Synthesis via Reactive Distillation. 2. Dynamic Simulation and Control Aspects," *Ind. Eng. Chem. Res.*, **36**, 1870–1881 (1997).
Walas, S. M. *Phase Equilibria in Chemical Engineering*, Boston: Butterworth-Heinemann (1985).

7

Other Unit Operations

7.1 Introduction

In the last several chapters we have discussed the control of the important unit operations of reactors, columns, and heat exchangers. In this chapter we briefly explore typical control structures and plantwide dynamic considerations for a variety of other unit operations. The treatment is at best sketchy and at worst superficial, but we hope it is sufficient to provide some appreciation of the types of control systems that are used for these processes. More material of this type can be found in Shinskey (1988).

7.2 Furnaces

Fired furnaces are frequently used in chemical plants to provide energy at high temperatures. If a column, a reactor, or some other unit requires energy at a temperature level above that attainable by steam at reasonable pressure levels, a fired reboiler or heater is used. Steam in a chemical plant is seldom available for process use at pressures above 300 psia. The saturation temperature of 300 psia steam is 417°F, so if the column base temperature is above about 350°F, steam cannot be used.

A furnace is really a chemical reactor. The two reactants are fuel and air. The kinetics are almost instantaneous, and the heat of the exothermic reaction is high. All sorts of fuels are used: natural gas, waste gases from the plant, fuel oil, coal, wood, etc. Anything that will burn can be used. In fact some of the more interesting control problems occur in "trash-to-steam" plants burning municipal waste. The plant gets paid to take the trash and gets paid for the electrical energy it puts into the grid. This sounds like a great deal, but the tremendous variability of the trash fed to the furnace results in extreme variations in steam and power generation rates. Sometimes the trash contains a lot of water, and little heat is generated by its combustion. Other times the trash may contain items like old tires that give off large amounts of energy. So these trash-to-steam plants are subjected to very large and random load disturbances.

Figure 7.1 shows a typical furnace in which fuel and air are introduced into the combustion zone. A control valve is shown on the air line, but this is just schematic. More typically the manipulation of the air flow is achieved by varying the speed of a *forced-draft* fan (on the inlet of the furnace). The pressure inside the furnace is controlled by varying the speed of an *induced-draft* fan (on the outlet of the furnace).

Furnaces usually have to deal with "on-demand" load changes, i.e., a customer instantaneously needs more steam or more heat input to a unit. The control system on the furnace must be set up to respond quickly to these load changes. The process shown in Fig. 7.1 shows a furnace in which a stream is being heated in a furnace. The outlet temperature of the process stream is controlled by adjusting the fuel flowrate. The air flow is ratioed to the fuel flow. This ratio is adjusted by the output signal from an excess oxygen controller that looks at the composition of the stack gas. The use of too much air increases energy consumption, but too little air can lead to air pollution problems due to incomplete combustion.

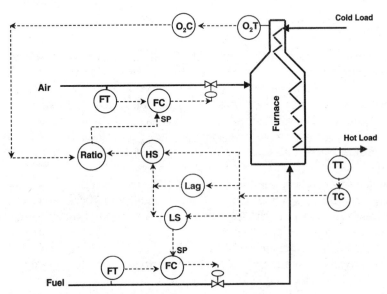

Figure 7.1 Furnace firing controls.

An interesting aspect of furnace control is the need to be always on the air-rich side, never on the fuel-rich side. If the furnace became filled with uncombusted fuel and then air was added, the resulting rapid combustion could blow the furnace apart. The same concern makes it important that the start-up of a furnace follow a very carefully thought-out procedure. The control system shown in Fig. 7.1 accomplishes this air-rich operation by the use of several selectors and a lag unit. When the temperature controller calls for more fuel, the air will increase first before the fuel increases because the low selector on the fuel passes the low signal from the lag to the fuel flow controller while the high selector on the air passes the high signal to the air flow controller. The reverse operation occurs when the temperature controller calls for less fuel: The fuel flow decreases first and then the air flow decreases.

7.3   Compressors

There are three important types of compressors commonly used in chemical plants: reciprocating, axial, and centrifugal. Control of reciprocating compressors involves varying either speed (strokes per minute) or displacement (length of each stroke). Control of centrifugal and axial compressors involves two important aspects: manipulation of throughput and antisurge control. We discuss each of these below. Compressors are driven by electric motors or steam turbines. Synchronous electric motors are usually used, and these devices run at constant speed (fixed revolutions per minute). Turbines operate at different speeds and are more energy-efficient in many applications because of this ability to vary speed and save power.

7.3.1   Throughput control

There are three basic ways to control throughput in a centifugal compressor: spillback (bypassing), suction throttling, and variable speed. Figure 7.2 shows each of these structures and illustrates how they work in terms of their compressor curves.

Figure 7.2a shows a spillback setup in which a stream is recycled from the compressor discharge back to the suction of the compressor. Note that the recycle flow should come from downstream of a cooler to prevent a buildup in temperature due to the work of compression if a large fraction of the flow through the compressor is recycled. The compressor is driven by an electric motor operating at constant speed. Let us assume that the suction and discharge pressures are constant. Let the flowrate of the stream entering and leaving the system be F and the flowrate through the compressor be F_{comp}. The flowrate through the recycle line is F_{bypass}. The total flow through the compressor F_{comp} is

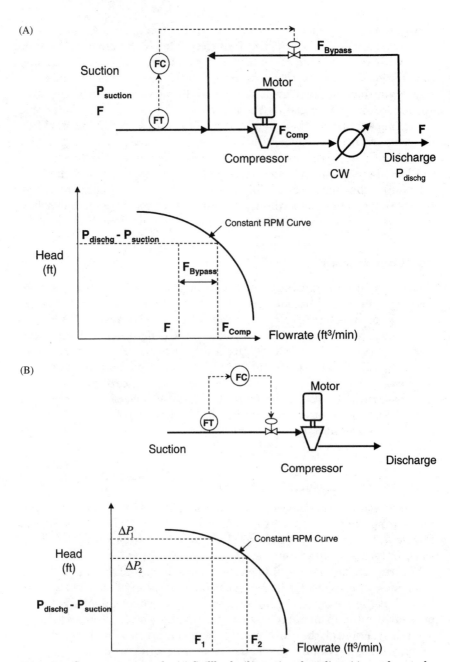

Figure 7.2 Compressor controls. (*a*) Spillback; (*b*) suction throttling; (*c*) speed control.

(C)

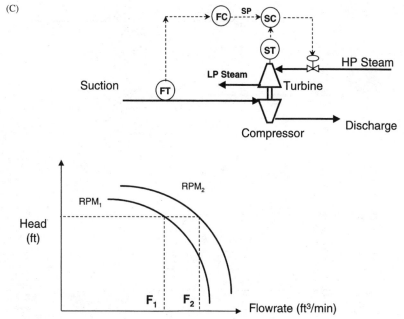

Figure 7.2   *(Continued)*

constant no matter what the flowrate to and from the system is. The recycle flowrate F_{bypass} is changed by opening or closing the bypass valve to control F. Thus the compressor sees a constant flowrate and a constant head, and therefore the power consumption is constant. The compressor operates at a fixed point on the compressor curve. Reducing throughput results in no reduction in electric power demand.

Figure 7.2b shows a suction-throttling configuration in which the control valve in the suction of the compressor is adjusted to change the throughput. Now the compressor moves up and down on its compressor curve. Less throughput (going from F_2 to F_1) is achieved by pinching the suction valve, which drops the suction pressure and raises the head required of the compressor ($P_{\text{dischg}} - P_{\text{suction}}$ increases ΔP_2 to ΔP_1). This backs the compressor up on its curve, giving less flow. Although the head increases, the power decreases because of the reduction in flow. So suction throttling is more energy-efficient than spillback.

Figure 7.2c illustrates how a variable-speed drive (a steam turbine) can be used to control throughput. The turbine is driven by high-pressure steam (600 psia) and discharges into a low-pressure steam header (25 psia). A flow control/speed control cascade structure is used. The output signal from the flow controller adjusts the setpoint of the turbine speed controller, which manipulates the flow of high-pressure

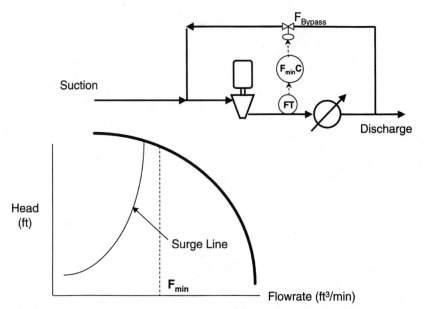

Figure 7.3 Compressor antisurge control.

steam to the turbine. Now the compressor moves horizontally on its compressor curves to change throughput. The variable-speed strategy is the most energy-efficient.

7.3.2 Antisurge control

Centrifugal compressors can go unstable if the flowrate of gas through the compressor is reduced too far. To avoid this, an antisurge controller is used that prevents the flowrate through the compressor from dropping below some minimum. Recycling from compressor discharge to compressor suction is used to keep the load on the compressor (see Fig. 7.3).

7.4 Decanters

Since there are two liquid phases in a decanter, the total liquid level and the interface level must both be controlled. Of course this requires that they both be measured. Measuring the interface between a liquid and a vapor is usually not difficult because the density difference between the two phases is large. However the density difference between two liquid phases is relatively small, and this can make liquid-liquid interface measurement difficult. Figure 7.4a shows a conventional control system for a decanter. The light phase is removed to hold the top level, and the heavy phase is removed to hold the liquid-liquid interface.

Figure 7.4b shows a decanter control scheme proposed many years ago by Page Buckley. The idea is to recognize that the level of the light liquid phase is affected by the flowrates of both the light and heavy streams leaving the decanter. On the other hand, the interface level is only affected by the flowrate of the heavy liquid stream leaving the decanter. The scheme uses a flow controller to control the sum of the two flowrates by manipulating the light stream, and adjusting the setpoint of this total flow controller by the output signal from the top level controller. This eliminates the interaction between the two level loops.

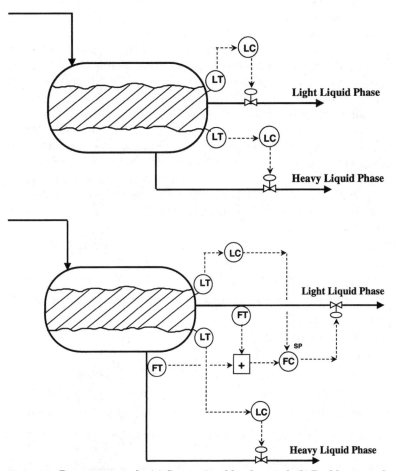

Figure 7.4 Decanter controls. (a) Conventional level control; (b) Buckley control structure to eliminate interaction.

7.5 Refrigeration Systems

When a process requires temperatures lower than those attainable by cooling water (less than 100°F), refrigeration must be used. Most refrigeration systems use compression refrigeration, but occasionally absorption refrigeration systems are employed if an inexpensive low-level heat source is available.

Figure 7.5 shows two typical compression refrigeration systems. The first uses a fixed-speed motor to drive the refrigeration compressor, so compressor suction throttling is used. The temperature of the hot stream to be cooled is held by using a temperature/pressure cascade scheme. The pressure of the boiling refrigerant is controlled by manipulating the vapor valve. The temperature controller adjusts the setpoint of the pressure controller. Liquid refrigerant is brought into the evaporator to hold liquid level.

When the compressor is driven by a variable-speed turbine, the temperature controller adjusts the setpoint of the turbine speed controller, which manipulates the flowrate of high-pressure steam to the turbine.

Figure 7.6 gives a control scheme for an absorption refrigeration plant in which ammonia is used as the refrigerant. Ammonia is produced as the distillate product from a distillation column operating at high pressure so that cooling water can be used in the condenser. The liquid ammonia is flashed into an evaporator operating at low pressure, and the boiling ammonia cools the process stream. The vapor ammonia is absorbed in a "weak liquor" water stream, and the "strong liquor" mixture is fed into the distillation column to separate the water and ammonia. The control scheme contains the following loops:

1. The temperature of the chilled process stream is controlled by adjusting the setpoint of the evaporator level controller (the heat transfer area is varied to change the heat transfer rate).

2. The water stream used to absorb the ammonia vapor is ratioed to the ammonia flowrate.

3. The level in the base of the column is controlled by manipulating the feed flowrate to the column.

4. The temperature profile is controlled by manipulating heat input to the reboiler.

5. Reflux drum level is controlled by manipulating reflux flowrate.

6. Column pressure is controlled by cooling water flow to the condenser.

Note that the liquid level in the strong liquor surge tank is not controlled. The water and ammonia circulate around in the system, but nothing leaves the system except for very small losses.

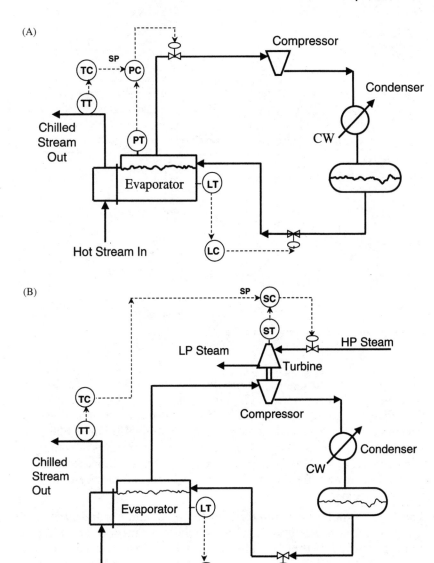

Figure 7.5 Compression refrigeration controls. (*a*) Fixed-speed compressor; (*b*) variable-speed compressor.

7.6 Plant Power Utility Systems

A number of utilities are required to run a chemical plant: cooling water, steam at various pressure levels, electricity, refrigeration, air, and inert gas (nitrogen). Those utilities associated with providing power (electricity and steam) normally generate both electricity and process steam. The two most common power plants are:

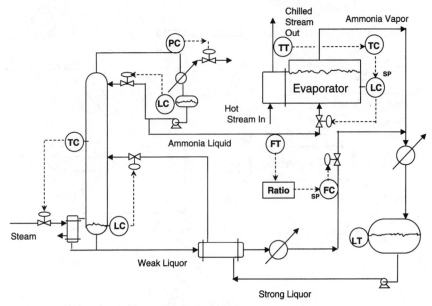

Figure 7.6 Absorption refrigeration controls.

- *Fired boiler generating high-pressure (1000 psia) steam:* The steam is generated and superheated in a furnace. It flows through a turbine that drives a generator producing electrical power. Steam flows at various pressures are extracted from the turbine. Steam at 600 psia may be used to drive steam turbines connected to pumps or compressors. Steam at 300 psia may be used as a high-temperature heat source. Steam at 150 psia and 25 psia may be used for lower-temperature requirements.

- *Combustion turbine driving a generator:* Fuel and air are burned in a turbine. The hot combustion gases drive the turbine. The hot gas from the turbine is then used to generate low-pressure steam for use in the plant. Combustion turbines are more efficient than furnaces, but they require relatively clean fuel.

After the steam is used in the various plant unit operations, it is typically returned as condensate to the boiler house for reuse.

Figure 7.7 shows a typical furnace/boiler configuration and its control system. Steam pressure in the 1000-psia header is maintained by furnace firing (the air controls are not shown). Steam pressure in the 300-psia header is controlled by either removing more steam from the turbine if the pressure is too low or dropping steam into the 150-psia header if the pressure is too high. Similar strategies are used for the

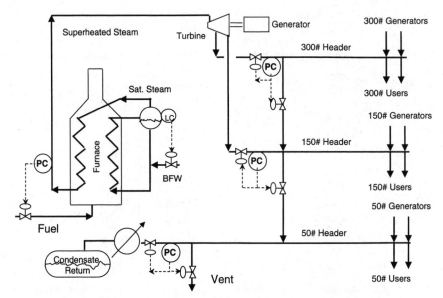

Figure 7.7   Plant power utility system controls.

other two steam headers. Because utilities are required on the demand of downstream units, their control systems are set up to handle rate changes quickly.

7.7   Liquid-Liquid Extractors

Liquid-liquid extraction is a separation unit used to transfer a substance from one liquid phase to another liquid phase. The unit exploits the difference in solubility between these partially miscible phases.

Two liquid phases exist in a liquid-liquid extractor, and either the light or the heavy phase can be continuous (droplets of the discontinuous phase rising or falling through the continuous phase). Different types of contacting devices are used: trays, packing, agitated stages, etc.

The pressure and the interface level must be controlled in a liquid-liquid extraction column. We assume that the column is running liquid-full (no vapor phase at the top of the column). If the light phase is continuous, the liquid-liquid interface is in the base of the column. Interface level is controlled by the removal rate of the heavy phase, and pressure is controlled by the flowrate of the light phase out the top of the column. If the heavy phase is the continuous phase, the interface is at the top of the column and it is controlled by adjusting the flowrate of the light phase (Fig. 7.8).

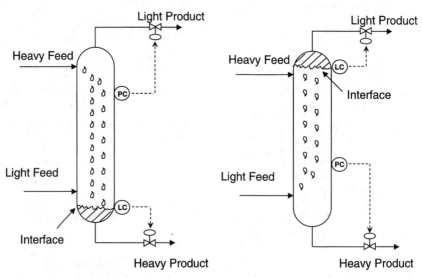

Figure 7.8   Liquid-liquid extractor controls.

7.8   Multiple-Effect Evaporators

Figure 7.9 shows two types of multiple-effect evaporators. In the first, both the fresh feed and the steam are fed to the first stage. In the second, the fresh feed is fed to the first effect and the steam is fed to the last stage. The reasons for using one flowsheet or the other include temperature sensitivity of the process material and the pressure level of the available steam. Multiple-effect evaporators are used to conserve energy. Operating each stage at different pressures allows the vapor from one stage to be used as the heating medium in the next.

The levels in each evaporator are controlled by the liquid stream leaving the vessel. The temperature in the last stage is controlled by manipulating the steam flowrate, even if the steam is not used in the reboiler at this stage but is added at the other end of the evaporator series.

7.9   Conclusion

We have presented a very brief discussion of typical control schemes for a number of important unit operations: furnaces, compressors, decanters, steam/power processes, liquid-liquid extractors, and evaporators. Each of these units requires the control of certain key variables if it is to do its job. A basic regulatory control system must be in place on each unit.

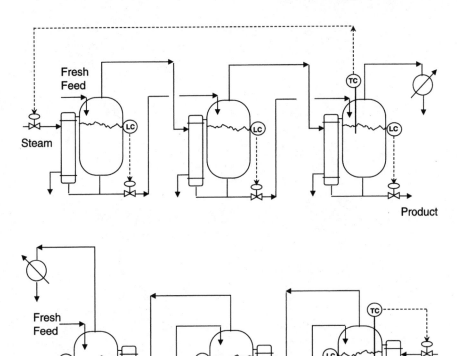

Steam

Product

Figure 7.9 Multiple-effect evaporator controls. (*a*) Energy flow in same direction as process flow; (*b*) energy flow in direction opposite process flow.

However, the control system for all unit operations must be consistent with the plantwide requirements of the coupled and interconnected process. These can sometimes dictate control structures for units that appear abnormal if only a narrow unit-operation perspective is considered. Units do not function in isolation, and their control systems must reflect this interconnectivity.

7.10 Reference

Shinskey, F. G. *Process Control Systems*, 3d ed., New York: McGraw-Hill (1988).

PART

3

Industrial Examples

8

Eastman Process

8.1 Introduction

Now that we have laid the groundwork by looking at the control of individual unit operations, we are ready to return to the plantwide control problem. In the next four chapters we illustrate the application of the nine-step design procedure with four industrial process examples.

We begin with a fairly simple process consisting of a reactor, condenser, separator, compressor, and stripper with a gas recycle stream (Fig. 8.1). This process was developed and published by Downs and Vogel (1993) as an industrial plantwide control test problem. A FORTRAN program is available from them that does the derivative evaluations for the process. The user must write a main program that initializes the simulation, does the controller calculations, performs the numerical integration, and plots the results.

A detailed description of the process in this book is unnecessary since one was provided in the original paper. We summarize here only some of the essential and unusual dynamic features. A small amount of an inert noncondensible component B is introduced in a feed stream and must be purged from the process. There are four fresh gas feed streams: F_{oA}, F_{oD}, F_{oE}, and F_{oC}. The first three are mixed with the recycle gas and fed into the bottom of the reactor. The last fresh feed F_{oC} is fed into the bottom of the stripper.

There are two main reactions, both of which are irreversible and exothermic:

$$A + C + D \rightarrow G \tag{8.1}$$

$$A + C + E \rightarrow H \tag{8.2}$$

Two additional irreversible and exothermic side reactions produce by-

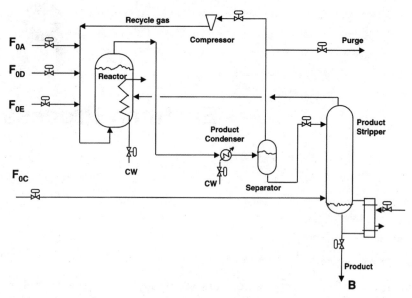

Figure 8.1   Eastman process flowsheet and nomenclature.

product F. The reactions are approximately first-order with respect to reactant concentrations. The reactor, which is open-loop unstable, contains both liquid and vapor phases, but no liquid stream leaves the reactor. Vapor from the reactor flows through a partial condenser and into a separator drum. Liquid from the drum is fed to the top tray of a stripping column. Vapor from the drum is compressed, a small portion is purged, and the remainder is recycled back to the reactor. The stripper has two sources of vapor: a small reboiler and the F_{oC} fresh feed.

Both gas pressure and liquid level in the reactor are integrating phenomena, and the choice of manipulated variables to control them is somewhat clouded. Temperature, pressure, and liquid level in the reactor all interact and their behavior is nonlinear. The gas purge stream from the process is very small, so its effectiveness in controlling pressure is doubtful.

The four fresh reactant feed streams must be managed in an appropriate way to satisfy overall component balances. Fortunately, composition analyzers are available. Figure 8.2 gives a sketch of the process with nomenclature and the values of flowrates, compositions, temperatures, and pressures at the initial steady state (Mode 1).

Several different control structures have been published in the literature for the Eastman process: Ricker (1993), McAvoy and Ye (1994), Price et al. (1994), Lyman and Georgakis (1995), Ricker and Lee (1995), Banerjee and Arkun (1995), Kanadibhotla and Riggs (1995), McAvoy

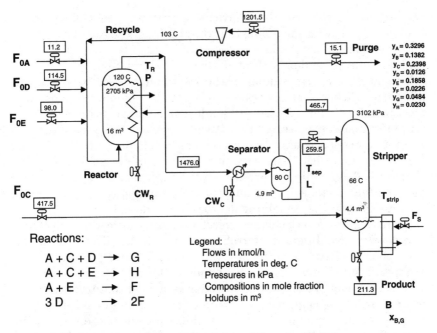

Figure 8.2 Eastman process base steady-state conditions for Mode 1.

et al. (1996), and Ricker (1996). Several common loops appear in nearly all of these control strategies. Reactor temperature is typically controlled with cooling water to the reactor, but in some structures the setpoint of this reactor temperature controller is changed via a cascade configuration. Most of the published structures control liquid levels in the separator and stripper by manipulating the liquid streams leaving those vessels. The fresh feed F_{oA} is generally used to control the composition of component A in the system and the purge to control component B composition.

However, many differences arise among the schemes when looking at how production rate is set and how liquid level and pressure in the reactor are controlled. For example, production rate is set in various strategies via fresh feed F_{oD} flow, condenser cooling water flow, separator liquid flow, stripper base flow, or fresh feed F_{oC} flow. Reactor level is controlled by fresh feeds F_{oD} and F_{oE}, separator temperature setpoint, compressor recycle valve, or fresh feed F_{oC} flow. Reactor pressure is controlled by reactor cooling water flow, purge flow, or F_{oA} feed flow. In one strategy reactor pressure is uncontrolled and allowed to float.

Throughout this book we have repeatedly stressed that the plantwide control problem is open-ended, which means there is no unique correct solution. For this process it is also not clear where production rate needs to be set, so researchers have chosen different control objectives. Hence it is not at all surprising that the studies of the Eastman process have yielded different control structures. Their performance can be evaluated in the end only through dynamic simulation.

The most extensive studies of the Eastman process have been reported by Ricker and coworkers in a number of publications. In a most insightful work, Ricker (1996) developed a decentralized control structure and demonstrated that it was superior to the more complex nonlinear model predictive control scheme he had presented previously. He also gives a lucid discussion of the issues in developing a control structure: determining degrees of freedom, selecting variables that must be controlled, selecting how to set production rate, and deciding what to do with the remaining degrees of freedom.

In the next two sections, we select two different control objectives and develop two different control structures. In the first case, we assume that the flowrate of the product stream B leaving the base of the stripper is set by a downstream customer. Hence it is flow-controlled, and the setpoint of the flow controller is a load disturbance to the process. In the second case, we assume that the fresh feed stream F_{oC} is set by an upstream process. So it is flow-controlled, and the setpoint of the flow controller is a load disturbance to the process. These two cases demonstrate that different control objectives produce different control strategies.

8.2   Case 1: On-Demand Product

8.2.1   Regulatory control strategy

Step 1.   We are assuming in this section that the product stream from the bottom of the stripper is set on the demand of a downstream user. The bottoms stream from the stripper is flow-controlled and so we set the position of the control valve, XMV(8), on this stream (B). The rest of the liquid level controls must be chosen to accommodate this first-priority choice. Note that we could put a flow controller on this stream if necessary, but this was not done in the simulations described later. The quality specification is that component G in the product should not vary more than ± 5 mol %.

Step 2.   This process has 12 degrees of freedom. One of these is agitation rate, which we simply hold constant. This leaves 11 degrees of freedom: four fresh feeds F_{oA}, F_{oD}, F_{oE}, and F_{oC}; purge valve; gas recycle valve; separator base valve; stripper base valve; steam valve; reactor cooling water valve; and condenser cooling water valve.

Step 3. The open-loop instability of the reactor acts somewhat like a constraint, since closed-loop control of reactor temperature is required. By design, the exothermic reactor heat is removed via cooling water in the reactor and product condenser. We choose to control reactor temperature with reactor cooling water flow because of its direct effect. There are no process-to-process heat exchangers and no heat integration in this process. Disturbances can then be rejected to the plant utility system via cooling water or steam.

Step 4. Because of the objective to achieve on-demand production rate, the product stream leaving the stripper base is flow-controlled via the bottoms control valve. This is a good example of how a degree of freedom must be used to satisfy a design or business constraint.

Step 5. There is only one product, and the only quality specification is that the composition of component G should not vary more than ± 5 mol %. In most processes there would be a specification on the amount of component E permitted in bottoms product from the stripper. However, the problem statement makes no mention of controlling the impurity E in the product stream. The manipulator of choice to control product quality $(x_{B,E})$ is stripper steam flow (F_S) because of its fast response. Stripper temperature can be used to infer product composition. It does a good job in keeping most of the light components from being lost in the product. There are only small changes in $x_{B,E}$ for the disturbances specified by Downs and Vogel (1993) (± 0.5 percent). Another manipulated variable that directly affects stripper bottoms purity is the flowrate of feed F_{oC}. However, this fresh feed makeup stream affects the component balances of A and C in the system, while steam does not, and we would have recognized this at Step 7. Therefore, we choose reboiler steam to control product purity.

High reactor temperature (175°C) is one safety constraint. Reactor cooling water flow has previously been selected to control reactor temperature. The only other known safety constraint for this process is pressure, which must not exceed the shut-down limit of 3000 kPa. The gas fresh feed streams, cooling water streams, reboiler steam flow, and purge directly affect pressure. Any of these could be used to control pressure. Of course the reaction rate also affects pressure, so a variable that changes the reaction rate could potentially be used to control pressure indirectly. Reactor cooling water flow and steam have already been selected. Condenser cooling rate is smaller than reactor cooling rate, so it may not be very effective in controlling pressure. This leaves one of the gas flows.

The purge stream is only 15.1 kmol/h, while the largest fresh feed makeup stream, F_{oC}, is 417 kmol/h. The vapor holdup in the reactor, separator, and stripper is estimated to be about 15 m³. This gives a time constant of about 2 minutes if F_{oC} is used to control pressure. If the purge flow is used, the time constant is about 60 minutes. Because fresh feed F_{oC} is the largest of the gas flows by far, we choose it to control pressure.

Step 6. Three liquid levels need to be controlled: reactor, separator, and stripper base. We must use the Buckley strategy of *level control in the reverse direction to flow* since the stripper base product B is fixed by production rate. Therefore liquid flow from the separator (L) must be used to control stripper base level. To control level in the separator, we select the cooling water flow to the condenser (CW_C).

Now we must decide how to control the liquid level in the reactor. This liquid consists of mostly the heavy products, components G and H. The more fresh reactant components D and E are fed into the process, the more products will be produced. So we select the two fresh feed flowrates F_{oD} and F_{oE} to control reactor liquid level. We ratio one to the other depending upon the desired split between components G and H in the final product. Simple flow ratios should be accurate enough to maintain the desired product distribution without any feedback of product compositions. So on-line analyzers on the product streams should not be required.

Step 7. A light inert component B enters in one of the feed streams. It can be removed from the process only via the purge stream, so purge flowrate is used to control the composition y_B in the purge gas stream. Stripper temperature control keeps the volatile gas reactants within the gas recycle loop. Components D and E are accounted for via reactor level. The component balance for C is maintained via pressure, assuming we can control the composition of the other major component A in the gas loop. There must be some feedback mechanism to guarantee that precisely the correct number of moles of this component are fed into the system to react with the number of moles of component C. The only manipulator available to satisfy the component balance for A is the fresh feed stream F_{oA}. So we select this flow to control the composition of A in the purge gas stream y_A. The compositions of either the purge gas or the reactor feed could be used, but both are not necessary.

Only two analyzers are required to run this process: one measuring the amount of A in the system and one measuring the amount of B. The latter could be eliminated if the amount of inert B coming into the system does not vary drastically. The purge stream is very small, and variations in the concentration of B in the system should have only a minor effect on controllability. However, Ricker (1996) has shown that the purge does have a significant economic impact.

Step 8. Control of the individual unit operations has been established and all of the control valves have been assigned except for the gas recycle valve.

Step 9. Of the original 11 degrees of freedom, we have used one for production rate, one for reactor temperature control, one for product quality, one for pressure control, three for liquid levels, and two for compositions. An additional one was used to set the G/H ratio. This leaves one degree of freedom to be specified. This is the valve that controls the flowrate of the gas recycle. We fix this valve to be wide open, based on the Douglas heuristic (Fisher et al., 1988) that gas recycle flows should be maximized to improve yields.

We have used the reactor cooling water valve to stabilize the system by controlling reactor temperature. However there is no specific temperature at which the reactor must operate. The best way to manage the reactor temperature setpoint is not immediately obvious. It might be used in conjunction with the production rate controller, i.e., higher temperatures may be needed to increase throughputs. It might be adjusted to maximize yields and suppress undesired by-products.

However, after making some simulation runs with several of the disturbances suggested in the original paper, it became apparent that the temperature in the separator was changing quite substantially and adversely affecting the stripper. Low separator temperature drops too many light components into the stripper, and the reboiler steam has trouble maintaining product quality. Therefore a separator temperature controller was added, whose output signal is the setpoint of the reactor temperature controller. The final basic regulatory control structure (Fig. 8.3) is simple, effective, and easily understood by operating personnel.

8.2.2 Override controls

The basic regulatory control structure outlined above was able to hold the process at the desired operating point for most of the disturbances. However, when manipulated variables hit constraints it was unable to prevent a unit shutdown. The disturbance IDV(6) that shuts off the fresh feed flowrate F_{oA} is probably the most drastic. The resulting imbalance in the stoichiometric amounts of components A and C drives the concentration y_A down quite rapidly. The reaction rate slows up, reactor temperature drops, and the process shuts down on high pressure. Since one degree of freedom has been removed by this disturbance, the control structure must be modified with overrides to handle the component balances.

The F_{oC} stream contains more C than A, so the excess C must be

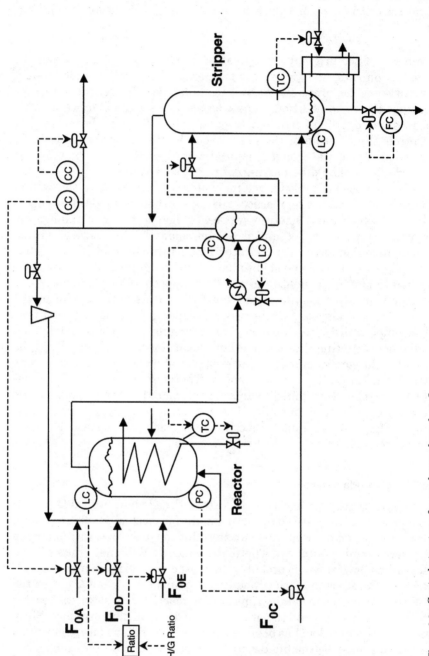

Figure 8.3 Control structure for on-demand product flow.

removed from the system. The only place available is the purge stream. Therefore, a low F_{oA} flow override controller is used to open the purge valve (Fig. 8.4). The other action that must be taken is to prevent the concentration of component A in the system from dropping down too low and reducing reaction rates. This is achieved by using a low y_A concentration override controller to pinch the fresh feed flowrate F_{oD} to slow up the rate of consumption of A. Of course F_{oE} is also reduced through the ratio.

Now the liquid level loops must also be modified, since we no longer can specify production rate and reactor level control cannot use F_{oD}. This is easily accomplished by using low level override controllers on each of the three levels. Low stripper level pinches product base product flowrate B. Low separator level pinches separator liquid flowrate L. Low reactor level pinches the condenser cooling water flowrate CW_C. In an override situation the level control structure has been reversed from the basic structure and now levels are held *in the direction of flow*.

8.2.3 Simulation results

Dynamic simulations were run with the proposed control structure for all disturbances proposed by Downs and Vogel (1993). Only the Mode 1 operation was studied. Section 8.5 gives the FORTRAN program used. Figures 8.5 to 8.8 show results for several disturbances.

Figure 8.5 shows how large changes in production rate are handled. Product stream B *immediately* changes, and the rest of the process flows adjust appropriately. At time equals 1 hour, the valve position on the production rate (B) is dropped 50 percent. At time equals 10 hours, it is increased back to its base case value. The process follows these changes in B by gradually and smoothly reducing fresh feeds through the level controllers.

Figure 8.6 gives results for changing the G/H split in the product. The two fresh feed streams F_{oD} and F_{oE} are immediately changed to the appropriate new values. The reactor level controller output signal is sent to a flow controller on F_{oD}, and the bias value on this flow controller is changed to the desired value. At the same time the ratio between the two flowrates is set to the new desired number. Product composition $x_{B,G}$ and $x_{B,H}$ change to their new values in about 4 hours.

Figure 8.7 shows how the process rides through the loss of F_{oA}, disturbance IMV(6). The override controller takes action almost immediately, and production rate is reduced after about 5 hours when liquid levels drop.

Figure 8.8 shows the responses to IMV(1), a change in the composition of A and C in the F_{oC} stream. As the amount of A in the system drops, the override controller cuts feed streams F_{oD} and F_{oE}. When reactor

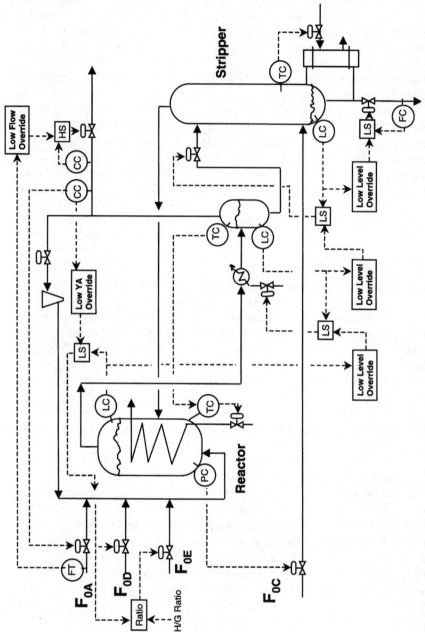

Figure 8.4 Control structure for on-demand product flow with overrides.

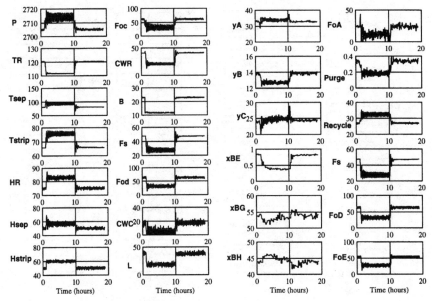

Figure 8.5 Dynamic response for 50 percent change in product B flow.

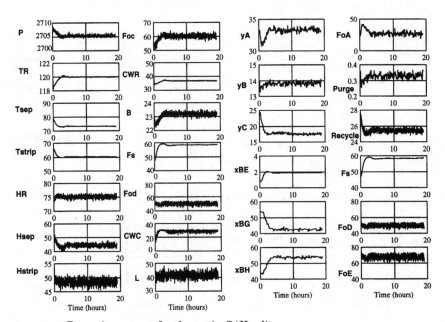

Figure 8.6 Dynamic response for change in G/H split.

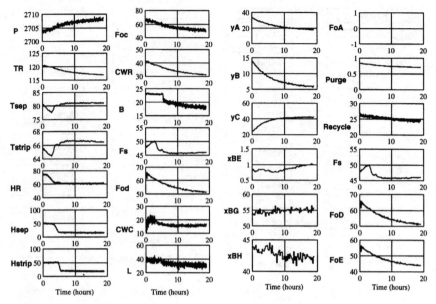

Figure 8.7 Dynamic response for loss of fresh feed F_{oA}.

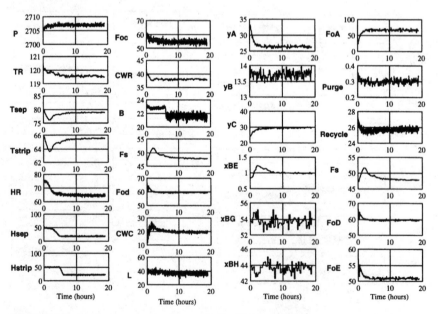

Figure 8.8 Dynamic response for change in F_{oC} feed composition.

TABLE 8.1 Controller Tuning Constants

		K_c	Transmitter span
	Basic Control Loops		
Levels	Reactor	4	100%
	Separator	2	100%
	Stripper	2	100%
Pressure	Reactor	100	3000 kPa
Temperatures	Reactor	3	100°C
	Separator	0.15	100°C
	Stripper	2	100°C
Compositions	y_B	16	100 mol %
	y_A	10	100 mol %
	Override Controllers		
Levels	Reactor	1	100%
	Separator	2	100%
	Stripper	2	100%
Composition	y_A	1	100 mol %
Flow	F_{oA}	100	100%

holdup drops, the override controller cuts condenser cooling. When separator level drops the override controller cuts separator liquid L. Finally, after about 25 hours, the low level in the stripper cuts back slightly on product rate B.

8.2.4 Controller tuning

A word needs to be said about controller type and controller tuning. Controller algorithm selection and tuning are important to the success of any control system. Two features should be recognized about the Eastman process. First, it is an integrating process with little self-regulation in terms of pressure, liquid levels, and chemical components. Second, there are no tight specifications on any variables.

The integrating nature of this process makes it difficult to tune controllers with integral action. Two integrators in series presents a challenging control problem because 180° of phase angle is lost. The absence of tight specifications implies that steady-state offset or error is not a problem. Thus both of these features lead us to use simple proportional-only controllers on all loops. Both the basic regulatory controllers and the overrride controllers are P-controllers.

Table 8.1 gives values for the controller gains used and the transmitter spans. All the valves are spanned in the provided program between 0 and 100 percent. The level controller gains ranged from 1 to 4 and required little tuning. The loops that required a little empirical tuning were the three temperatures, the pressure, and the two compositions.

Tuning was performed by increasing the controller gain and testing the dynamic response to a step change in setpoint until the loop became too oscillatory. Reactor temperature was tuned first, followed by pressure, separator temperature, stripper temperature, component A composition, and component B composition. No claim is made that these are the best settings, but they give adequate control and required little time to tune.

A plantwide control design procedure was used to develop a simple but effective regulatory control system for the Eastman process with an on-demand product control objective. With this strategy, control of production rate is essentially instantaneous. Drastic upsets and disturbances are handled by simple proportional-only overrides.

8.3   Case 2: On-Supply Reactant

8.3.1   Regulatory control strategy

Step 1.   In this section we alter the control objective relating to production rate. Instead of flow controlling the product stream from the bottom of the stripper, we assume that an upstream process sets the flow of the F_{oC} stream and the process must take whatever amount is fed into it. Most of the steps in the design procedure are the same as the previous section, but the control of liquid levels is now in the direction of flow. The product quality criterion is the same.

Step 2.   The same number of degrees of freedom exist.

Step 3.   This is constructed as above.

Step 4.   Production rate is set by fresh feed F_{oC}.

Step 5.   Reboiler steam controls product purity. Now the fresh feed F_{oC} cannot be used to control pressure. The purge stream is so small that effective pressure control is unlikely. Reactor cooling water flow is used to control reactor temperature. Therefore, the logical choice for pressure control is the cooling water flow to the condenser CW_C. The controller gain of this loop was empirically set at 25 (with a pressure transmitter span of 3000 kPa).

Step 6.   We use the Buckley strategy of *level control in the direction of flow* to regulate two liquid levels. The stripper base level is controlled by manipulating stripper bottoms flow. The separator level is controlled by manipulating the liquid flowrate from the separator to the stripper. As before, the liquid level in the reactor is controlled by the two fresh feed flowrates F_{oD} and F_{oE}. We ratio one to the other depending upon the desired split between components G and H in the final product.

Step 7. Purge flow controls the composition y_B in the recycle gas stream. The fresh feed F_{oA} controls the composition y_A in the recycle gas stream.

Step 8. All control valves have been assigned.

Step 9. Separator temperature is controlled by changing the setpoint of the reactor temperature controller. The controller gain of the separator temperature controller was empirically set at 0.5 (with a temperature transmitter span of 100°C).

8.3.2 Control scheme and simulation results

The control system is shown in Fig. 8.9. The override controller on low F_{oA} flow was used to open the purge valve. The override controller on low y_A composition was used to pinch the fresh feed flowrate F_{oD}.

Typical simulation results are shown in Fig. 8.10. At time equals zero, the fresh feed flowrate F_{oC} is reduced by 25 percent from its base-case value. The process responds to this change by gradually cutting back on the other feed streams and the product leaving the unit. The new steady-state conditions are attained in a little over 1 hour. The control structure also successfully handled the other disturbances.

8.4 Conclusion

Developing a plantwide control system for the Eastman process is fairly straightforward. There are only five unit operations and one gas recycle stream. No energy integration is present. So the major feature of this process from a plantwide viewpoint is the problem of accounting for the multiple component inventories.

We have shown how different control objectives lead to different control structures. Although the two strategies handle disturbances differently, they both work without showing a clear advantage of one compared with the other.

In the next three chapters, we consider more challenging processes with many more unit operations and multiple recycle streams.

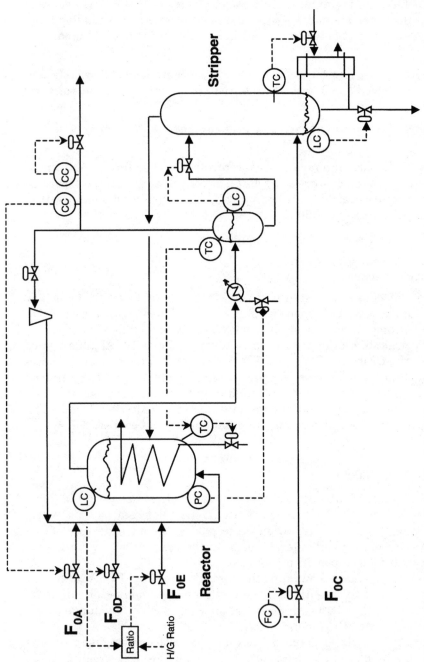

Figure 8.9 Control structure for fixed fresh feed F_{0C}.

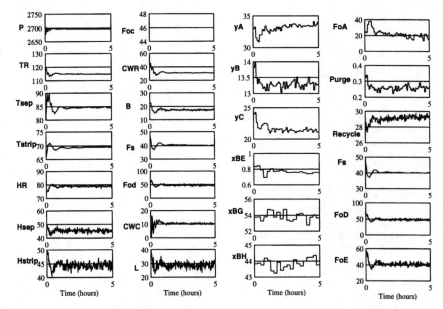

Figure 8.10 Dynamic response for 25 percent reduction in F_{oC} feed flow.

8.5   FORTRAN Program for Eastman Process

On-Demand Product

```
c
c Control structure 1: "eastcs1.for"
c Production rate (B) is flow controlled
c TC of Tsep added, changing TR
c Overrides added on pressure to purge
c                      stripper level to B
c                      separator level to L
c                      yA to FoD
c                      reactor level to CWC
      DOUBLE PRECISION XMEAS,XMV,SETPT
      COMMON/PV/XMEAS(41),XMV(12),SETPT(21)
      INTEGER IDV
      real kc1,kc2,kc3,kc4,kc5,kc6,kc7,kc8,kc9,kc10
      real lp(5000)
      COMMON/DVEC/IDV(20)
      double precision gain(21),reset(21)
      dimension trp(5000),prp(5000),hrp(5000),timep(5000)
      dimension tsepp(5000),hsepp(5000),tstripp(5000)
      dimension hstripp(5000),yap(5000),ybp(5000),ycp(5000)
      dimension fodp(5000),cwcp(5000),focp(5000)
      dimension cwrp(5000),foap(5000),purgep(5000)
      dimension foep(5000),bp(5000),fsp(5000)
      dimension xbep(5000), xbgp(5000), xbhp(5000)
      dimension recycp(5000),tlagp(5000)
```

```fortran
      double precision yy(51),yp(51),time
      open(10,file='wl4.dat')
      open(9,file='wl3.dat')
      open(8,file='wl2.dat')
      open(7,file='wl1.dat')
      data tprint,tplot/0.,0./
      delta=1./3600.


      nn=51
      call teinit(nn,time,yy,yp)
      do 10 k=1,20
 10   idv(k)=0

      write(6,2)time,xmeas(7),xmeas(8),xmeas(9)
 2    format(' time=',f6.3,' P=',f6.0,' HR=',f6.2,
     +  ' TR=',f6.2 )
      write(6,3)xmeas(11),xmeas(12)
 3    format('           Tsep=',f6.2,' Hsep=',f6.2)
      write(6,4)xmeas(29),xmv(10),xmeas(17)

 4    format('        yA=',f6.2,' Purge=',f6.2,' B=',f6.2)
      ip=0
      tplot=0.
      tstop=19.
c Specify disturbance number

      idv(1)=1
c****************************************************
c Disturbance is 15% reduction in production rate
c     xmv(8)=xmv(8)*0.85
c     xmv(8)=xmv(8)*1.2
c****************************************************
      xmv8o=xmv(8)
      xmv8base=xmv8o
c  Base case is fodo=63.053
      ratio=53.98/63.053
      fodo=63.053
c Disturbance is switch from 50/50 G/H to 1/3 G/H
c     fodo=50.434
c     ratio=68.38/50.434
c****************************************************
 100  continue
c Put in series of rate changes
c     if(time.gt.1.)xmv8o=xmv8base*0.5
c     if(time.gt.10.)xmv8o=xmv8base
      if(time.lt.tprint)go to 20
      write(6,2)time,xmeas(7),xmeas(8),xmeas(9)
      write(6,3)xmeas(11),xmeas(12)
      write(6,4)xmeas(29),xmeas(10),xmeas(17)
      tprint=tprint+.1
 20   if(time.lt.tplot)go to 21
      ip=qip+1
      timep(ip)=time
      trp(ip)=xmeas(9)
      prp(ip)=xmeas(7)
      hrp(ip)=xmeas(8)
      tsepp(ip)=xmeas(11)
```

```
         hsepp(ip)=xmeas(12)
         tstripp(ip)=xmeas(18)
         hstripp(ip)=xmeas(15)
         yap(ip)=xmeas(29)
         ybp(ip)=xmeas(30)
         ycp(ip)=xmeas(31)
         foap(ip)=xmv(3)
         if(idv(6).eq.1)foap(ip)=0.
         focp(ip)=xmv(4)
         fodp(ip)=xmv(1)
         cwcp(ip)=xmv(11)
         cwrp(ip)=xmv(10)
         foep(ip)=xmv(2)
         bp(ip)=xmeas(17)
         lp(ip)=xmv(7)
         fsp(ip)=xmv(9)
         xbep(ip)=xmeas(38)
         xbgp(ip)=xmeas(40)
         xbhp(ip)=xmeas(41)
         purgep(ip)=xmeas(10)
         recycp(ip)=xmeas(5)
         tlagp(ip)=tlag
         tplot=tplot+0.02
21       continue
c Level Control Loops
c LC 1: hr(8) controlled by fod(1)
         kc1=4.
         err1=75-xmeas(8)
         xmv(1)=fodo+kc1*err1
c*********************************************
c Low yA override pinches fod
         yakc=1.
         yaerr=30.-xmeas(29)
         yafod=fodo-yakc*yaerr
         if( yafod.lt.xmv(1))xmv(1)=yafod
c*****************************************************
c Ratio foe to fod
         xmv(2)=xmv(1)*ratio
c LC 2: hsep(12) controlled by CWc(11)
         kc2=2.
         err2=50.-xmeas(12)
         xmv(11)=18.114+kc2*err2
c*************************************************
c Low reactor level pinches condenser CWc
         hrkc=1.
         hrerr=75.-xmeas(8)
         hrcwc=30.-hrkc*hrerr
         if( hrcwc.lt.xmv(11))xmv(11)=hrcwc
         if(xmv(11).lt.0.)xmv(11)=0.
         if(xmv(11).gt.1000.)xmv(11)=100.

c*****************************************************
c*************************************************8
c Low stripper level pinches B
         xmv(8)=xmv8o
         bor=100.+2.*(xmeas(15)-50.)
         if( bor.lt.xmv(8))xmv(8)=bor
c*********************************************
```

```
c LC 3: hstrip(15) controlled by L(7)
      kc3=2.
      err3=50.-xmeas(15)
      xmv(7)=38.1+kc3*err3
c*********************************************
c Low separator level pinches L
      xlor=100.+2.*(xmeas(12)-50.)
      if( xlor.lt.xmv(7))xmv(7)=xlor
c*********************************************
c PC 4: pr(7) controlled by foc(4)
c Ramp up pset
c      pset=2705.+pramp*time
c       if(pset.gt.pmax)pset=pmax
c Disturbance in pset: 2705 to 2645.
      pset=2705.
      kc4=100.
      err4=(pset-xmeas(7))/30.
      xmv(4)=61.302+kc4*err4
      if(xmv(4).lt. 0.)xmv(4)=0.
c Temperature Control Loops
c Tsep (11) controlled by trset
      tsepkc=0.15

      tsepset=80.109

      tseperr=tsepset-xmeas(11)
      trset=120.4+tsepkc*tseperr
c TC 5: TR (9) controlled by CWR (10)
      kc5=3.
      err5=(trset-xmeas(9))
      xmv(10)=41.106-kc5*err5

c TC 6: Tstrip(18) controlled by Fs(9)
      tstrpset=65.731
      kc6=2.
      err6=tstrpset-xmeas(18)
      xmv(9)=47.446+kc6*err6
      if(xmv(9).lt.0.)xmv(9)=0.
      if(xmv(9).gt.100.)xmv(9)=100.


c Composition Control Loops
c CC 7: yA (23) controlled by Foa (3)
      kc7=10.
      err7=32.188-xmeas(23)
      xmv(3)=24.644+kc7*err7
      if(xmv(3).gt.100.)xmv(3)=100.
      if(xmv(3).lt.0.)xmv(3)=0.
c CC 8: yB (24) controlled by purge (6)
      kc8=16.
      err8=13.823-xmeas(30)
      xmv(6)=40.064-kc8*err8
c*********************************************
c Loss of FoA opens purge valve

      if(idv(6).eq.1)xmv(6)=100.
c*********************************************
c Integration
```

```
      call tefunc(nn,time,yy,yp)
      do 50 k=1,51
      yy(k)=yy(k)+yp(k)*delta
50       continue

      time=time+delta
      if(xmeas(7).gt.2950.)go to 89
      if(time.lt.tstop)go to 100
89 do 90 k=1,ip
      write(7,91)timep(k),trp(k),prp(k),hrp(k),tlagp(k)
      write(8,91)tsepp(k),hsepp(k),tstripp(k),hstripp(k)
   +    ,yap(k),ybp(k),ycp(k)
      write(9,91)foap(k),focp(k),fodp(k),cwcp(k),cwrp(k)
   +    ,recycp(k)
      write(10,91)foep(k),bp(k),lp(k),fsp(k),xbep(k)
   +    ,xbgp(k),xbhp(k),purgep(k)
91       format(8(1x,f12.5))
90       continue
      stop
      end
```

8.6 References

Banerjee, A., and Arkun, Y. "Control Configuration Design Applied to the Tennessee Eastman Plantwide Control Problem," *Comput. Chem. Eng.*, **19,** 453–480 (1995).

Downs, J. J., and Vogel, E. F. "A Plant-Wide Industrial Process Control Problem," *Comput. Chem. Eng.*, **17,** 245–255 (1993).

Fisher, W. R., Doherty, M. F., and Douglas, J. M. "The Interface Between Design and Control. 3. Selecting a Set of Controlled Variables," *Ind. Eng. Chem. Res.*, **27,** 611–615 (1988).

Kanadibhotla, R. S., and Riggs, J. B. "Nonlinear Model Based Control of a Recycle Reactor Process," *Comput. Chem. Eng.*, **19,** 933–948 (1995).

Lyman, P. R., and Georgakis, C. "Plantwide Control of the Tennessee Eastman Problem," *Comput. Chem. Eng.*, **19,** 321–331 (1995).

McAvoy, T. J., and Ye, N. "Base Control for the Tennessee Eastman Problem," *Comput. Chem. Eng.*, **18,** 383–413 (1994).

McAvoy, T. J., Ye, N., and Gang, C. "Nonlinear Inferential Parallel Cascade Control," *Ind. Eng. Chem. Res.*, **35,** 130–137 (1996).

Price, R. M., Lyman, P. R., and Georgakis, C. "Throughput Manipulation in Plantwide Control Structures," *Ind. Eng. Chem. Res.*, **33,** 1197–1207 (1994).

Ricker, N. L. "Model Predictive Control of a Continuous, Nonlinear, Two-Phase Reactor," *J. Proc. Cont.*, **3,** 109–123 (1993).

Ricker, N. L., and Lee, J. H. "Nonlinear Model Predictive Control of the Tennessee Eastman Challenge Process," *Comput. Chem. Eng.*, **19,** 961–981 (1995).

Ricker, N. L. "Decentralized Control of the Tennessee Eastman Challenge Process," *J. Proc. Cont.*, **6,** 205–221 (1996).

NOMENCLATURE

B = flow rate of stripper bottoms

CW_C = flow rate of cooling water to condenser

CW_R = flow rate of cooling water to reactor

F_{oj} = flow rate of fresh feed, $j = A, C, D, E$

F_S = flow rate of steam to stripper

HR = reactor holdup, percent level

H_{sep} = separator holdup, percent level

H_{strip} = stripper holdup, percent level
L = flow rate of liquid from separator to stripper
P = pressure
Purge = purge gas flow rate
Recycle = flow rate of gas recycle to reactor
TR = reactor temperature
T_{sep} = separator temperature
T_{strip} = stripper temperature
x_{Bj} = composition of stripper bottoms, mole fraction component j
y_j = composition of purge gas, mole fraction component j

CHAPTER

9

Isomerization Process

9.1 Introduction

In the previous chapter we studied a fairly simple process consisting basically of a boiling-liquid reactor and a simple separation section. Although the Eastman process has some plantwide control features, it is essentially just a nonlinear reactor control problem. The gas recycle loop acts like a big stirrer. The management of chemical components through fresh feed makeup streams and product streams is the principal aspect that illustrates plantwide control considerations.

In this and the next two chapters, we study more realistically complex chemical processes. These processes feature multiple unit operations connected together with recycle streams.

We begin here with a very simple isomerization process that is similar to some of the simplified processes studied in Chap. 2. The process consists of a reactor, two distillation columns, and a liquid recycle stream. There are four components to consider. In subsequent chapters we look at processes with many more units, many more components, and multiple recycle streams.

Figure 9.1 shows the flowsheet of the isomerization process to convert normal butane (nC_4) into isobutane (iC_4).

$$nC_4 \rightarrow iC_4 \tag{9.1}$$

This process is quite important in the petroleum industry because isobutane is usually more valuable as a chemical feedstock than normal butane. The typical amount of iC_4 contained in crude oil and produced in refinery operations such as catalytic cracking is sometimes not enough to satisfy the demand. On the other hand the supply of nC_4 sometimes exceeds the demand, particularly in the summer when less nC_4 can be blended into gasoline because of vapor pressure limitations.

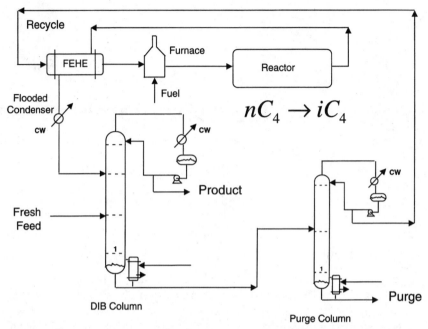

Figure 9.1 Isomerization process flowsheet.

Some of the many uses of isobutane include the production of high-octane gasoline blending components by reacting it with various olefins in alkylation processes and the production of propylene oxide and tertiary butyl alcohol.

The reaction of nC_4 to iC_4 occurs in the vapor phase and is run at elevated temperatures (400°F) and pressures (600 psia). The reaction is exothermic (heat of reaction -3600 Btu/lb · mol), so there is a temperature rise as the process stream flows through the adiabatic tubular reactor. Following heat exchange with the reactor inlet stream and condensation with cooling water, the reactor effluent is introduced into a large distillation column that separates the C_4's. The iso/normal separation is difficult because of the similar relative volatilities, so many trays (50) and a high reflux ratio (7) are required. For the design case considered, this column ends up being 16 feet in diameter. This column is called a *deisobutanizer* (DIB).

The fresh feed stream is a mixture of both nC_4 and iC_4 (with some propane and isopentane impurities). It is also introduced into the column, not directly into the reactor. It is fed at a lower tray in the column than the reactor effluent stream because the concentration of iC_4 in the fresh feed is lower. This enables the removal of some of the iC_4 and all of the C_3 in the fresh feed before sending the nC_4 to the reactor from the recycle stream. The ratio of the recycle flow to the fresh feed flow is about 1:2. The DIB column operates at 100 psia so that cooling water can be used in the condenser (reflux drum temperature is 124°F). The base temperature is 150°F, so low-pressure steam can be used.

The distillate product from the DIB is the isobutane product. It has a specification of 2 mol % nC_4. Since the fresh feed contains some propane, there is also some propane in the distillate product. All of the propane in the feed leaves the process in the distillate stream.

The bottoms from the DIB contains most of the nC_4, along with some iC_4 impurity and all of the heavy isopentane impurity. Since this heavy component will build up in the process unless it is removed, a second distillation column is used to purge out a small stream that contains the isopentane. Some nC_4 is lost in this purge stream. The purge column has 20 trays and is 6 ft in diameter. The distillate product from the second column is the recycle stream to the reactor, which is pumped up to the required pressure and sent through a feed-effluent heat exchanger and a furnace before entering the reactor in the vapor phase.

The numerical case studied is derived from a flowsheet given in Stanford Research Institute Report 91, "Isomerization of Paraffins for Gasoline." Since no kinetic information is given in this report, only reactor inlet and exit conditions, we will assume two different types of kinetics. In Case 1 we consider that the reaction is irreversible. An activation energy of 30,000 Btu/lb · mol is used, and the preexponential factor is adjusted to give the same conversion reported in the SRI report. In Case 2 we assume that the reaction is reversible. The equilibrium constant decreases with increasing temperature because the reaction is exothermic. We also increase the size of the reactor so that the effluent leaves essentially at chemical equilibrium.

These artificial kinetics are used so that a comparison can be made of processes with reversible and irreversible reactions. In particular we want to demonstrate that the effect of increasing reactor temperature is completely different in these two cases. With irreversible reactions, increasing temperature increases production rate. With reversible reactions, increasing temperature can produce a decrease in production rate. Figure 9.2 gives conditions at the Case 1 steady state. Table 9.1 gives stream data for both cases. Table 9.2 lists the process parameter values.

9.2   Plantwide Control Strategy

Step 1.   In this process we want to achieve the desired production rate and control the impurity of normal butane in the isobutane product at 2 mol %. Reactor pressure cannot exceed the design operating pressure

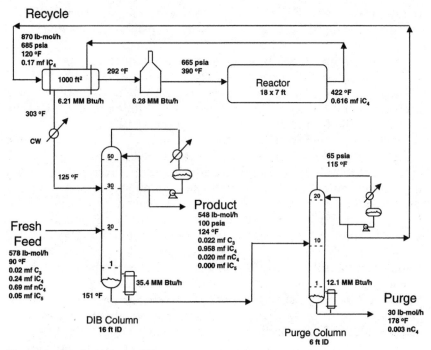

Recycle

870 lb-mol/h
685 psia
120 °F
0.17 mf iC$_4$

1000 ft^2

292 °F

6.21 MM Btu/h   6.28 MM Btu/h

665 psia
390 °F

Reactor
18 x 7 ft

422 °F
0.616 mf iC$_4$

303 °F

CW

125 °F

Fresh
Feed

578 lb-mol/h
90 °F
0.02 mf C$_3$
0.24 mf iC$_4$
0.69 mf nC$_4$
0.05 mf iC$_5$

50

30

20

1

151 °F

35.4 MM Btu/h

Product
548 lb-mol/h
100 psia
124 °F
0.022 mf C$_3$
0.958 mf iC$_4$
0.020 mf nC$_4$
0.000 mf iC$_5$

65 psia
115 °F

20

10

1   12.1 MM Btu/h

Purge

30 lb-mol/h
178 °F
0.003 nC$_4$

DIB Column
16 ft ID

Purge Column
6 ft ID

Figure 9.2 Steady-state conditions.

of 700 psia. We assume that we are free to choose the production rate
handle. Neither the fresh feed nor product flowrates are fixed by other
plant considerations. In the pentane purge column, we do not want to
lose too much nC_4.

Step 2. This process has 14 control degrees of freedom. They include:
fresh feed valve; DIB column steam, cooling water, reflux, distillate,
and bottoms valves; purge column steam, cooling water, reflux, distillate, and bottoms valves; furnace fuel valve; flooded condenser cooling
water valve; and DIB column feed valve.

Step 3. The exothermic heat of reaction must be removed, and the
reactor feed must be heated to a high enough temperature to initiate
the reaction. Since the heat of reaction is not large and complete one-
pass conversion is not achieved, the reactor exit temperature is only
32°F higher than the reactor inlet temperature. Since heat transfer
coefficients in gas-to-gas systems are typically quite low, this small
temperature differential would require a very large heat exchanger if
only the reactor effluent is used to heat the reactor feed and no furnace

TABLE 9.1   Stream Data

Case 1 (irreversible reaction)	Fresh feed	Product	Purge	Reactor inlet	Reactor outlet
Flow, lb · mol/h	580	550	30	870	870
Mole fraction C_3	0.02	0.021	0	0	0
Mole fraction iC_4	0.24	0.959	0	0.17	0.62
Mole fraction nC_4	0.69	0.020	0.01	0.81	0.36
Mole fraction iC_5	0.05	0	0.99	0.02	0.02
Temperature, °F	90	124	178	390	422
Pressure, psia	300	100	66	665	650

Cases 1 and 2	Furnace	Flooded condenser	DIB reboiler	Purge reboiler	Reflux flow
Duty, 10^6 Btu/h	6.3	7.3	35.4	12.1	
DIB, lb · mol/h					4000
Purge, lb · mol/h					700

Case 2 (reversible reaction)	Fresh feed	Product	Purge	Reactor inlet	Reactor outlet
Flow, lb · mol/h	544	514	30	870	870
Mole fraction C_3	0.02	0.021	0	0	0
Mole fraction iC_4	0.24	0.959	0	0.17	0.59
Mole fraction nC_4	0.69	0.020	0.01	0.82	0.40
Mole fraction iC_5	0.05	0	0.99	0.01	0.01
Temperature, °F	90	124	178	390	420
Pressure, psia	300	100	66		

is used. Therefore, a furnace is required to bring the reactor inlet temperature up to the desired level.

The use of a feed-effluent heat exchanger (FEHE) reduces the amount of fuel burned in the furnace, as discussed in detail in Chap. 5. So from a steady-state viewpoint, the economic trade-off between utility and capital costs would produce a fairly large heat exchanger and a small furnace. However, the exothermic heat of reaction and the heat of vaporization supplied in the furnace must be dissipated to utilities at the flooded condenser. If the FEHE is too large, reactor heat will be recycled. Also, the larger the heat exchanger, the smaller the heat input in the furnace. This means that we have a smaller handle to deal with disturbances. This could potentially be solved using a bypass around the FEHE on the cold side as discussed in Chap. 5. This should enable us to prevent reactor runaway to high temperature and would guarantee that the furnace is in operation at all times. However, unless the furnace is large enough, there is no guarantee that the system will never quench to low temperature when a large disturbance occurs to drop the reactor inlet temperature.

TABLE 9.2 Parameter Values

		Case 1	Case 2
Reactor	ID, ft	7	7
	Length, ft	18	50
	Holdup, ft^3	693	1925
FEHE	Area, ft^2	1000	1000
	Shell holdup, ft^3	21	21
	Tube holdup, ft^3	21	21
DIB	ID, ft	16	16
	Total trays	50	50
	Feed trays	20/30	20/30
	Reflux ratio	7.3	7.8
	Reflux drum holdup, ft^3	1700	1700
	Base holdup, ft^3	2000	2000
Purge column	ID, ft	6	6
	Total trays	20	20
	Feed tray	11	11
	Reflux ratio	0.8	0.8
	Reflux drum holdup, ft^3	370	370
	Base holdup, ft^3	400	400
Flooded Condenser	Holdup, ft^3	340	340
Kinetics	A_F, 1/h	4×10^8	4×10^8
	E_F, Btu/lb · mol	3×10^4	3×10^4
$K_{EQ} = e^{(A+B/T)}$	A		-10
	B, °R		5000

A second difficulty that can occur with a large FEHE is a hydraulic problem. The recycle stream entering the heat exchanger is a subcooled liquid (115°F) at the pressure in the reactor section (685 psia). As this stream is heated to the required reactor inlet temperature of 390°F, it begins to vaporize. It is superheated vapor when fed into the reactor (390°F at 665 psia). When a small FEHE is used, the exit temperature is 292°F for the recycle stream, which means the stream is still all liquid. All the vaporization occurs in the furnace. If a large heat exchanger were used, vaporization would begin to occur in the heat exchanger. This would make the hydraulic design of this FEHE much more difficult. The dynamic response could also be adversely affected as changes in flowrates and temperature make the stream go in and out of the two-phase region.

The same problem occurs on the hot side of the FEHE. As the hot reactor effluent is cooled, it starts to condense at some temperature. And this dewpoint temperature could occur in the heat exchanger and not in the flooded condenser if a large area is used.

Because of both the heat dissipation and hydraulic concerns, we use a relatively small FEHE: 1000 ft^2 compared to the Stanford Research Report's listed area of 3100 ft^2. So the energy management system consists of controlling reactor inlet temperature by furnace firing and controlling the rate of heat removal in the flooded condenser by cooling water flowrate. The heat of reaction and the heat added in the furnace are therefore removed in the flooded condenser. Because of this design we do not need a bypass around the FEHE.

Step 4. We are not constrained either by reactant supply or product demand to set production rate at a certain point in the process. We need to examine which variables affect reactor productivity.

The kinetic expression for the isomerization reaction is relatively simple. For the irreversible case, reaction rate depends upon the forward rate constant, reactor volume, and normal butane concentration $\mathcal{R} = k_F V_R C_{nC_4}$. From this expression we see that only three variables could possibly be dominant: temperature, pressure, and mole fraction of nC_4 in the reactor feed.

Pressure affects productivity through its influence on the reactant concentration. Since the normal operating reactor pressure is close to the design limit, we are constrained in how much we can move pressure to achieve the desired production rate change. The nC_4 mole fraction in the reactor feed is about 0.81. Therefore large absolute changes in the reactant feed mole fraction would have to be made to achieve a significant relative change in throughput.

Finally, we are then left with temperature. We see from Eq. (4.7) how the relative change in reaction rate depends upon the temperature through the activation energy. In this case the reaction rate increases by 20 percent for a 10°F change in temperature. Clearly temperature is a dominant variable for reactor productivity.

For the reversible case, reaction rate depends upon the forward and reverse rate constants, reactor volume, and nC_4 and iC_4 concentrations:

$$\mathcal{R} = k_F V_R C_{nC_4} - k_R V_R C_{iC_4}$$

The activation energy of the reverse reaction is always greater than the activation energy of the forward reaction since the reaction is exothermic. Therefore the reverse reaction will increase more quickly with an increase in temperature than will the forward reaction. Temperature may still dominate for reactor productivity, but in the opposite direction compared with the irreversible case, since conversion increases with lower temperature. However, when the temperature becomes too low, both reaction rates slow down such that we cannot achieve the desired production rate with this variable alone. Instead, the concentrations of nC_4 and iC_4 dominate the rate through the relationship imposed by the equilibrium constant $K_{eq} = C_{iC_4}/C_{nC_4}$.

Therefore we choose the reactor inlet temperature setpoint as the production rate handle for the irreversible case. However, for the reversible case we need to look for variables that affect the ratio of nC_4 to iC_4 in the recycle stream. For this case we will not have unit control for the reactor since these concentrations depend upon operation in other parts of the process.

Note that setting the production rate with variables at the reactor or within the process specifies the amount of fresh reactant feed flow required at steady state. The choices for the control system made in Steps 6 and 7 must recognize this relationship between production rate and fresh reactant feed flowrate.

Step 5.   The final isobutane product is the distillate from the DIB column, and we want to keep the composition of the nC_4 impurity at 2 mol %. Nothing can be done about the propane impurity. Whatever propane is in the fresh feed must leave in the product stream. Because the separation involves two isomers, the temperature profile is flat in the DIB column. Use of an overhead composition analyzer is necessary.

The choices of manipulated variables that can be used to control nC_4 composition in the DIB distillate include reflux flowrate, distillate flowrate, and reboiler heat input. Because the reflux ratio is high, control of reflux drum level using distillate flow may be ineffective, particularly if the distillate were going directly to a downstream process. If we decide to use reflux flow to control reflux drum level, we must control distillate composition by manipulating the distillate flowrate. The reason is that distillate flow must match production rate, which is independently set in the reactor. However, in this case we assume that the distillate is going to a storage tank or cavern, so large changes in distillate flowrate are not important. Distillate can then be used for reflux drum level control, allowing us to consider other variables for composition control.

Most distillation columns respond more quickly to vapor rate changes than to changes in liquid rates. Therefore, we select reboiler heat input to control nC_4 impurity in the distillate. This may seem like a poor choice because we are controlling something at the top of the column by changing a variable near the base. However, vapor changes affect all trays in the column quite quickly, so tight control of distillate composition should be possible by manipulating vapor boilup. A viable alternative is to control distillate composition with distillate flowrate and control reflux drum level with reflux flowrate.

For the reversible case, we are interested in the composition of the bottoms stream from the DIB column. We may consider dual composition control (controlling both the nC_4 impurity in the top and the iC_4 impurity in the bottom). Logic would dictate that the distillate composition should be controlled by manipulating the distillate flowrate and bottoms composition should be controlled by manipulating heat input.

To avoid the high-pressure safety constraint, we must control reactor pressure. We can use the distillate valve from the purge column, the flooded condenser cooling water valve, or the DIB column feed valve. The most logical variable to use for control of the flooded condenser (reactor) pressure is the DIB column feed valve (as shown in Fig. 5.5). Based upon the discussion in Step 3, we would then use the flooded condenser cooling water valve to keep the liquid level in a good control range.

Step 6. We have only two choices, DIB column base valve or purge column distillate valve, for fixing a flow in the recycle loop. Either of these would work. The rationale for picking one is based upon avoiding disturbances to the unit downstream of the fixed flow location. Since the purge column is not critical from the viewpoint of product quality, we elect to fix the flow upstream of reactor (purge column distillate flow) so that we minimize disturbances in reactor temperature and pressure.

We must control the two column pressures. As discussed in Chap. 6, this is best done by manipulating the condenser cooling water flowrates.

There are four liquid levels to be controlled. DIB column reflux drum level is controlled by manipulating distillate product flowrate. We must also control the levels in the DIB column base and in the purge column reflux drum and base.

Having made the choice to fix the purge column distillate flow, we are faced with the problem of how to control purge column reflux drum level. We have two primary choices: reflux flow or heat input. We choose the latter because the flowrate of the purge column reflux is small relative to the vapor coming overhead from the top of the column. Remember the Richardson rule, which says we select the largest stream.

The flowrate of the purge stream from the base of the purge column is quite small, so it would not do a good job in controlling base level. This is especially true when the large steam flow has been selected to control the reflux drum level. Base level in the purge column can, however, be controlled by manipulating the bottoms flowrate from the DIB column.

We are then left with controlling base level in the DIB column. The only remaining valve is the fresh nC_4 feed flowrate into the column. The feed is liquid and there are only 20 trays between the lower feed point and the column base, so base level control using feed should be possible. This base level is also an indication of the nC_4 inventory within the process.

The material balance control structure works opposite to the direction of flow. Purge column distillate is fixed; purge column reflux drum level is controlled by vapor boilup; purge column base level is controlled by the feed to the purge column; and DIB column base level is controlled by the fresh feed to the DIB column.

Had we started to assign the DIB column base level control first, we would have ended up with the same inventory control structure. The reason is as follows. Assume we had chosen the DIB column base valve to control base level. After resolving the purge column inventory loops, we would have found that we needed to control the purge column base or reflux drum level with the fresh feed flow to the DIB column. The dynamic lags associated with these loops would have forced us back to the control strategy as described above.

An obvious question at this point is "Why don't we just flow control the fresh feed into the process?" If we did this, we could not fix the flowrate in the recycle loop. For example, suppose we select the following control structure: fix fresh feed flowrate, control DIB column base level with DIB bottoms, control purge column base level with heat input, and control purge column reflux drum level with distillate. This structure is intuitively attractive and permits us to fix the production rate directly by setting the fresh feed flowrate. However, only level controllers set the flows around the recycle loop, so we would expect problems with snowballing. Flow disturbances can propagate around the liquid recycle loop. Simulation results given later in this chapter demonstrate that this is indeed what occurs with this structure. It does not give stable regulatory control of the process.

Step 7. Four components need to be accounted for. The light inert propane leaves in the product stream. The heavy inert component isopentane (iC_5) leaves in the purge stream. Any iC_4 coming into the process in the fresh feed and the iC_4 produced by the reaction can leave in the product stream.

The only component that is trapped inside the system and must be consumed by the reaction is nC_4. The composition controller on the DIB distillate stream permits only a small amount (2 mol %) of nC_4 to leave in the product stream. The purge stream from the bottom of the purge column permits only a small amount of nC_4 to escape. This purge stream can be simply flow-controlled if we don't mind losing a small amount of nC_4 with the iC_5 purge. If the amount of iC_5 in the fresh feed is small, this may be the simplest strategy and may have little economic penalty. Alternatively we could control the amount of nC_4 in the purge column bottoms by manipulating bottoms flowrate. Since there is a fairly large temperature change in the purge column, controlling the temperature on a suitable tray (tray 2) may be more practical since it eliminates the need for an on-line analyzer. In the simulations given later in this chapter, we adopt the simple strategy of flow-controlling the purge stream.

The amount of reactant nC_4 fed into the system must somehow be exactly balanced by the amount of nC_4 converted to product iC_4. The process acts almost like a pure integrator in terms of the moles of nC_4. The way this balancing of nC_4 is accomplished in the control structure shown in Figure 9.3 is by using the level in the base of the DIB column to indicate if the nC_4 is building up in the system or is being depleted. The material in the DIB base is mostly nC_4. There is a little iC_4 (16 percent) and a little iC_5 (5 percent), and the remainder is nC_4. So DIB base level changes reflect changes in nC_4 inventory in the process. If this level is increasing, fresh feed flowrate should be reduced because we are not consuming all the reactant being fed into the process. If the level is decreasing, fresh feed flowrate should be increased because we are consuming more reactant than we are feeding.

Table 9.3 summarizes the component balance control strategy.

Step 8. The previous steps have left us at this point with two unassigned control valves, which are the reflux flows to each column. As discussed in Chap. 6, these are independent variables and can be fixed by flow controllers. We do not need dual composition control for the irreversible case because only one end of both columns is a product stream leaving the process. These two reflux flowrates are available in Step 9 to use as optimizing variables or to improve dynamic response. However, we may need dual composition control in the DIB column for the reversible case as mentioned in Steps 4 and 5.

Step 9. When we use reactor inlet temperature for production rate control (irreversible case), the only remaining degrees of freedom for optimization are the reflux flows for the two columns and the setpoint of the distillate flowrate from the purge column (recycle flow).

In the reversible case, when the base composition of the DIB column or the recycle flowrate is used for production rate control, the remaining degrees of freedom are purge column reflux flow and the reactor inlet temperature.

9.3 Dynamic Simulations

The simulation of this process was constructed using the commercial simulator from Hyprotech Ltd. A copy of the case "isom.hsc" can be obtained from the Web page of William L. Luyben at Lehigh University at the following site: *wll0@lehigh.edu*.

In the HYSYS simulation, the flooded condenser is simulated as a cooler with a temperature control loop manipulating cooling water rate. The reactor inlet pressure is set by the exit of the liquid recycle pump on the purge column distillate stream.

All level controllers are proportional only with a gain of 2. Tempera-

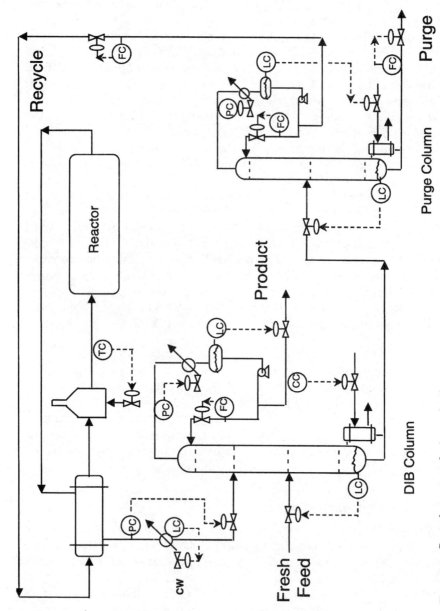

Figure 9.3   Control structure for isomerization process.

TABLE 9.3 Component Material Balance for Irreversible Case

Component	Input	+Generation	−Output	−Consumption	=Accumulation
					Inventory controlled by
C_3	Fresh feed	0	Product stream	0	Self-regulating by product quality controller
iC_4	Fresh feed	$k_F V_R C_{nC4}$	Product stream	0	Controlled by product quality controller
nC_4	Fresh feed	0	0	$k_F V_R C_{nC4}$	Indicated by DIB column base level
iC_5	Fresh feed	0	Purge stream	0	Self-regulating by composition change in purge

ture and composition controllers are tuned using the TL settings after the ultimate gain and frequency are obtained via relay-feedback tests. Two temperature measurement lags of 0.1 minute are included in the two temperature loops (reactor inlet temperature and DIB feed temperature). A 3-minute deadtime is assumed in the product composition measurement (distillate from the DIB). Table 9.4 gives valve and transmitter spans and controller tuning constants.

Rigorous dynamic simulations of the irreversible reaction case and the reversible reaction case permit us to evaluate how effective the

TABLE 9.4 Control Parameters

Reactor inlet temperature controller	Valve span, 10^6 Btu/h	15
	Transmitter span, °F	200
	K_c	2
	Reset, minutes	10
Cooler temperature controller	Valve span, 10^6 Btu/h	20
	Transmitter span, °F	100
	K_c	1
	Reset, minutes	10
Temperature measurement lags (two)	Time constant, minutes	0.1
Product composition controller	Valve span, 10^6 Btu/h	50
	Transmitter span, mole fraction nC_4	0.1
	K_c	1
	Reset, minutes	60
Composition deadtime	minutes	3
Level controllers	K_c	2

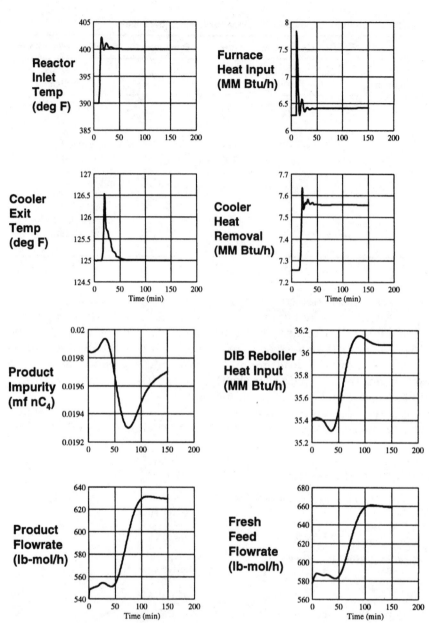

Figure 9.4 Irreversible reaction case dynamic response to 10°F increase in reactor inlet temperature.

control structure is. These simulations also alert us to any potentially difficult dynamics that would indicate tuning problems or a need to provide more surge capacity for some of the inventory loops. As we expected, the simulations show that the control strategy works and provides good base-level regulatory control.

9.3.1 Irreversible reaction

Figures 9.4 and 9.5 show results for the irreversible reaction case when reactor inlet temperature is changed and when recycle flowrate is changed. Figure 9.4 shows that a 10°F step increase, occurring at time equals 10 minutes, produces an increase in product flowrate (from 550 to 630 lb · mol/h). As more iC_4 is produced in the reactor with the increase in reaction rate at the higher temperature, the purity of the DIB distillate improves (less nC_4 in the overhead product), which causes the reboiler heat input to increase. This lowers the base level, and pulls in more fresh feed. So changing reactor inlet temperature is an effective way to set production rate when the reaction is irreversible.

Figure 9.5 shows the effect of increasing the recycle flowrate from 870 to 1000 lb · mol/h. If no other change is made, the effect is a slight *decrease* in production rate! This is certainly not what we would intuitively expect. This unusual behavior can be explained by consider-

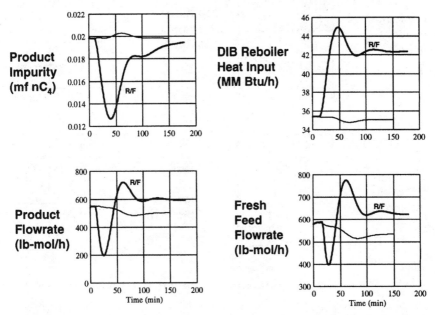

Figure 9.5 Irreversible reaction case dynamic response to increase in recycle flow from 870 to 1000 lb · mol/h (R/F curves indicate constant reflux-to-feed ratio in DIB column).

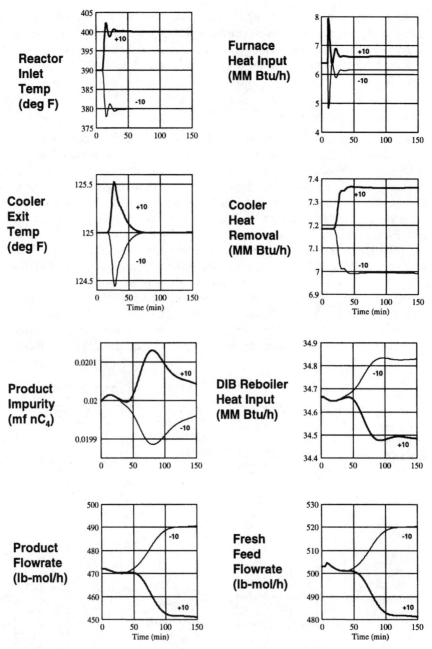

Figure 9.6 Reversible reaction case dynamic response to reactor inlet temperature change of +10 and −10°F.

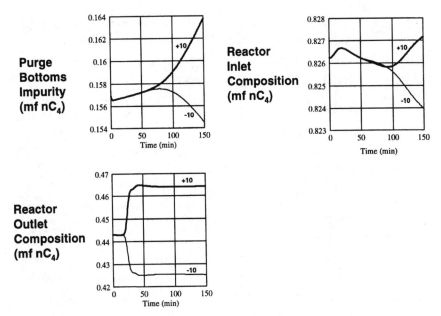

Figure 9.6 *(Continued)*

ing what happens in the DIB column. More recycle means a higher feed rate to this column, which increases the column load. Since we are holding the purity of the distillate, the variability is all reflected in the bottoms stream. Thus there is an increase in the iC_4 impurity in the bottoms. This changes the reactor inlet composition to have a lower reactant (nC_4) concentration. The result is a slight drop in the overall reaction rate.

The second set of curves labeled R/F in Fig. 9.5 shows what happens if a ratio controller is used to increase the reflux flowrate in the DIB column as the recycle flowrate is increased. Now the fractionating capability of the column increases as the load is increased, so the bottoms does not contain less nC_4 reactant. Product rate increases from 550 to 592 lb · mol/h.

We could also change production rate by keeping the recycle flowrate constant and only changing DIB reflux flowrate. This changes the bottoms product purity and affects the reactor inlet composition. We also could use dual composition control in the DIB column, in which case reflux is automatically adjusted.

9.3.2 Reversible reaction

Figures 9.6 and 9.7 give simulation results for the reversible reaction case where a large reactor is used so that the reaction is essentially

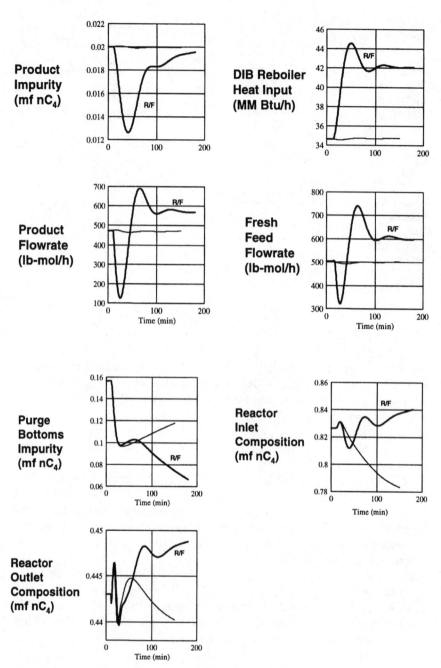

Figure 9.7 Reversible reaction case dynamic response to recycle flow increase (R/F curves indicate constant reflux-to-feed ratio in DIB column).

at equilibrium conditions in the reactor effluent. Figure 9.6 verifies that an *increase* in reactor inlet temperature results in a *decrease* in throughput and that a *decrease* in temperature *increases* throughput. This occurs because we have deliberately selected the kinetic parameters to give a decrease in the equilibrium constant as temperature increases, such as would occur with exothermic reversible reactions.

Note the changes in purge bottoms composition shown in Figure 9.6c. Remember we are using the simple strategy of flow controlling the purge stream. So the losses of nC_4 in the purge will vary with the fresh feed flowrate. An increase in fresh feed flowrate results in a decrease in purge nC_4 concentration because nC_4 is being replaced by the additional iC_5 coming in with the fresh feed stream. These iC_5 changes take many hours to occur because of the small amounts of iC_5 in the feed.

As we discussed in Step 4, the reversible case requires that we change the composition of the recycle stream to affect production rate. This can be done either by increasing the recycle to fresh feed ratio at constant fractionation in the DIB column or by affecting the fractionation in the DIB column at constant recycle flow. Figure 9.7 shows what happens when recycle flowrate is increased, with and without the reflux-to-feed ratio controller. When only the recycle flowrate is changed, production rate is essentially constant because we are effectively lowering the fractionation in the DIB column. More material is flowing through the reactor, and we would expect to get more iC_4 produced. But the change in the loading of the DIB column shifts the reactor inlet composition (less reactant nC_4). However, when we hold constant the reflux-to-feed ratio, an increase in recycle flow increases production rate.

9.3.3 Fixed fresh feed control structure

Figure 9.8 presents a control structure that most engineers would probably come up with for this process. Each of the units in isolation has standard "textbook" control strategies. Production rate is set by flow-controlling fresh feed to the DIB column. Liquid flows around the recycle loop are set by controlling levels. DIB base level is controlled by manipulating DIB bottoms flow. Reboiler heat input is used to control purge column base level. Purge column reflux drum level is controlled by manipulating distillate flowrate (recycle).

Figure 9.9 shows what happens for the irreversible reaction case. No disturbance is made. The system goes unstable and shuts down when control valve saturation occurs after about 2 hours. The reason for this should not come as a surprise at this point in the book. This "obvious" control strategy has no way to regulate the inventory of nC_4 component within the system.

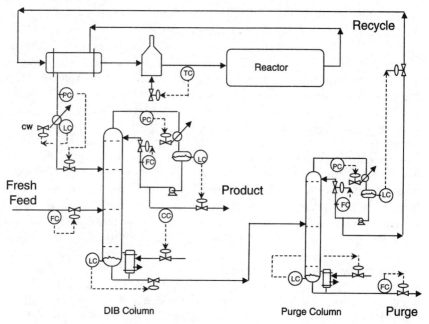

Figure 9.8 Fixed feed control structure.

9.4 Conclusion

The process considered in this chapter is very simple. The reaction involves only one reactant and one product. Two inert components, one light and one heavy, are also present. These inerts must be purged from the system. The major plantwide control consideration is how to adjust the fresh feed of the reactant to balance exactly its rate of consumption by reaction. This is achieved by using the liquid level that is a good indication of the amount of reactant in the system (the base level in the DIB column).

Despite its simplicity, the isomerization process displays some interesting behavior, some expected and some unexpected. The effect of recycle flowrate is probably the most nonintuitive phenomenon: recycle flowrate has little effect on throughput. This results because recycle flowrate affects both the reactor and the separation section, and the net effect is little change in production rate.

In the next two chapters, more complex flowsheets are considered. But many of the fundamental ideas developed in this chapter and in Chap. 2 are directly applicable to these complex, multiunit, multirecycle, multicomponent processes.

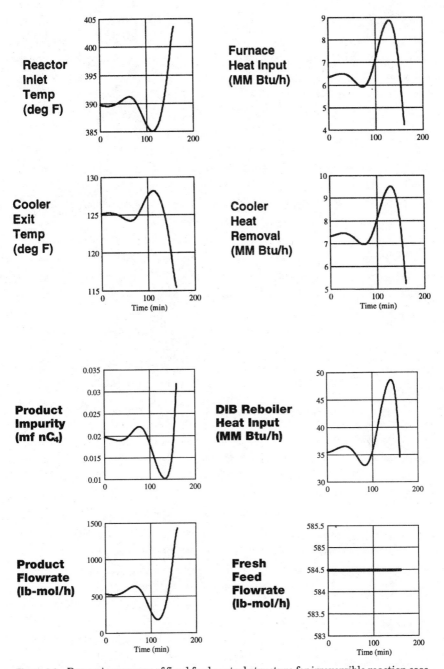

Figure 9.9 Dynamic response of fixed feed control structure for irreversible reaction case.

10

HDA Process

10.1 Introduction

We first presented the process for the hydrodealkylation (HDA) of toluene in Chap. 1 as part of our explanation of what the term *plantwide process control* means. We now return to it as the third illustration of our plantwide control design procedure.

Figure 10.1 shows the nine basic unit operations of the HDA process as described in Douglas (1988): reactor, furnace, vapor-liquid separator, recycle compressor, two heat exchangers, and three distillation columns. Two raw materials, hydrogen and toluene, are converted into the benzene product, with methane and diphenyl produced as by-products. The two vapor-phase reactions are

$$\text{Toluene} + \text{H}_2 \rightarrow \text{benzene} + \text{CH}_4 \qquad (10.1)$$

$$2\text{Benzene} \rightleftharpoons \text{diphenyl} + \text{H}_2 \qquad (10.2)$$

The kinetic rate expressions are functions of the partial pressures (in psia) of toluene p_T, hydrogen p_H, benzene p_B, and diphenyl p_D, with an Arrhenius temperature dependence. Zimmerman and York (1964) provided the following rate expressions:

$$r_1 = 3.6858 \times 10^6 \exp\left(-25{,}616/T\right) p_T \, p_H^{1/2} \qquad (10.3)$$

$$r_2 = 5.987 \times 10^4 \exp\left(-25{,}616/T\right) p_B^2$$
$$- 2.553 \times 10^5 \exp\left(-25{,}616/T\right) p_D \, p_H \quad (10.4)$$

where r_1 and r_2 have units of lb · mol/(min · ft³) and T is the absolute temperature in kelvin. The heats of reaction given by Douglas (1988) are $-21{,}500$ Btu/lb · mol of toluene for r_1 and 0 Btu/lb · mol for r_2.

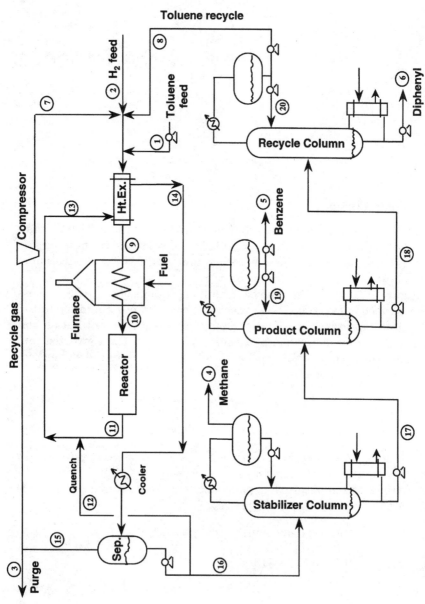

Figure 10.1 HDA process flowsheet.

The effluent from the adiabatic reactor is quenched with liquid from the separator. This quenched stream is the hot-side feed to the process-to-process heat exchanger, where the cold stream is the reactor feed stream prior to the furnace. The reactor effluent is then cooled with cooling water, and the vapor (hydrogen, methane) and liquid (benzene, toluene, diphenyl) are separated. The vapor stream from the separator is split. Part is purged from the process to remove the methane by-product and the remainder is sent to the compressor for recycle back to the reactor.

The liquid stream from the separator (after part is taken for the quench) is fed to the stabilizer column, which has a partial condenser and removes any remaining hydrogen and methane gas from the liquid components. The bottoms stream from the stabilizer is fed to the product column, where the distillate is the benzene product from the process and the bottoms is toluene and diphenyl fed to the recycle column. The distillate from the recycle column is toluene that is recycled back to the reactor and the bottoms is the diphenyl byproduct.

Makeup toluene liquid and hydrogen gas are added to both the gas and toluene recycle streams. This combined stream is the cold-side feed to the process-to-process heat exchanger. The cold-side exit stream is then heated further up to the required reactor inlet temperature in the furnace, where heat is supplied via combustion of fuel.

Pure component physical property data for the five species in our simulation of the HDA process were obtained from *Chemical Engineering* (1975) (liquid densities, heat capacities, vapor pressures, etc.). Vapor-liquid equilibrium behavior was assumed to be ideal. Much of the flowsheet and equipment design information was extracted from Douglas (1988). We have also determined certain design and control variables (e.g., column feed locations, temperature control trays, overhead receiver and column base liquid holdups) that are not specified by Douglas. Tables 10.1 to 10.4 contain data for selected process streams. These data come from our TMODS dynamic simulation and not from a commercial steady-state simulation package. The corresponding stream numbers are shown in Fig. 10.1. In our simulation, the stabilizer column is modeled as a component splitter and tank. A heater is used to raise the temperature of the liquid feed stream to the product column. Table 10.5 presents equipment data and Table 10.6 compiles the heat transfer rates within process equipment.

10.2 Plantwide Control Strategy

Step 1. For this process, as always, we must be able to achieve a specified production rate of essentially pure benzene while minimizing

TABLE 10.1 Process Stream Data, Part 1

	Fresh toluene	Fresh hydrogen	Purge gas	Stabilizer gas	Benzene product	Diphenyl product
Stream number	1	2	3	4	5	6
Flow, lb · mol/h	290.86	490.38	480.88	21.05	272.5	6.759
Temperature, °F	86	86	115	113	211	559
Pressure, psia	575	575	480	480	30	31
H_2, mole fraction	0	0.97	0.3992	0	0	0
CH_4	0	0.03	0.5937	0.9349	0	0
C_6H_6	0	0	0.0065	0.0651	0.9997	0
C_7H_8	1	0	0.0006	0	0.0003	0.00026
$C_{12}H_{10}$	0	0	0	0	0	0.99974

TABLE 10.2 Process Stream Data, Part 2

	Gas recycle	Toluene recycle	Furnace inlet	Reactor inlet	Reactor effluent	Quench
Stream number	7	8	9	10	11	12
Flow, lb · mol/h	3519.2	82.14	4382.5	4382.5	4382.5	156.02
Temperature, °F	115	272	1106	1150	1263.2	113
Pressure, psia	513	30	513	503	486	486
H_2, mole fraction	0.3992	0	0.4291	0.4291	0.3644	0
CH_4	0.5937	0	0.4800	0.4800	0.5463	0.0515
C_6H_6	0.0065	0.00061	0.0053	0.0053	0.0685	0.7159
C_7H_8	0.0006	0.99937	0.0856	0.0856	0.0193	0.2149
$C_{12}H_{10}$	0	0.00002	0	0	0.0015	0.0177

TABLE 10.3 Process Stream Data, Part 3

	FEHE hot in	FEHE hot out	Separator gas out	Stabilizer feed	Stabilizer bottoms	Product bottoms
Stream number	13	14	15	16	17	18
Flow, lb · mol/h	4538.5	4538.5	4156.0	382.5	361.4	88.91
Temperature, °F	1150	337	113	113	200*	283
Pressure, psia	486	480	486	480	480	33
H_2, mole fraction	0.3518	0.3518	0.3992	0	0	0
CH_4	0.5294	0.5294	0.5937	0.0515	0	0
C_6H_6	0.0907	0.0907	0.0065	0.7159	0.7538	0.0006
C_7H_8	0.0260	0.0260	0.0006	0.2149	0.2275	0.9234
$C_{12}H_{10}$	0.0021	0.0021	0	0.0177	0.0187	0.0760

*Stream temperature not based upon bubble point but set by heater.

TABLE 10.4   Process Stream Data, Part 4

	Product column reflux	Recycle column reflux
Stream number	19	20
Flow, lb · mol/h	300.0	12.0
Temperature, °F	211	272
Pressure, psia	30	30
H_2, mole fraction	0	0
CH_4	0	0
C_6H_6	0.9997	0.00061
C_7H_8	0.0003	0.99937
$C_{12}H_{10}$	0	0.00002

TABLE 10.5   Equipment Data and Specifications

Reactor	Diameter	9.53 ft
	Length	57 ft
FEHE	Area	30000 ft^2
	Shell volume	500 ft^3
	Tube volume	500 ft^3
Furnace	Tube volume	300 ft^3
Separator	Liquid volume	40 ft^3
Product column	Total theoretical trays	27
	Feed tray	15
	Diameter	5 ft
	Theoretical tray holdup	2.1 lb · mol
	Efficiency	50%
	Reflux drum liquid holdup	25 ft^3
	Column base liquid holdup	30 ft^3
Recycle column	Total theoretical trays	7
	Feed tray	5
	Diameter	3 ft
	Theoretical tray holdup	1 lb · mol
	Efficiency	30%
	Reflux drum liquid holdup	100 ft^3
	Column base liquid holdup	15 ft^3

TABLE 10.6   Heat Transfer Rates

FEHE	19.4 MW
Furnace	0.984 MW
Separator condenser	5.47 MW
Product reboiler	2.18 MW
Product condenser	2.05 MW
Recycle reboiler	0.439 MW
Recycle condenser	0.405 MW
Reactor heat generation	1.83 MW

yield losses of hydrogen and diphenyl. The reactor feed ratio of hydrogen to aromatics must be greater than 5:1. The reactor effluent gas must be quenched to 1150°F.

Step 2. There are 23 control degrees of freedom. They include: two fresh feed valves for hydrogen and toluene; purge valve; separator base and overhead valves; cooler cooling water valve; liquid quench valve; furnace fuel valve; stabilizer column steam, bottoms, reflux, cooling water, and vapor product valves; product column steam, bottoms, reflux, distillate, and cooling water valves; and recycle column steam, bottoms, reflux, distillate, and cooling water valves.

Step 3. The reactor operates adiabatically, so for a given reactor design the exit temperature depends upon the heat capacities of the reactor gases, reactor inlet temperature, and reactor conversion. Heat from the adiabatic reactor is carried in the effluent stream and is not removed from the process until it is dissipated to utility in the separator cooler. If we make the process-to-process heat exchanger too large, this means we may recycle the reactor heat and not dissipate it unless we have a bypass line around the exchanger. This is particularly true if the furnace is used only for trim control because it may not be able to provide sufficient control for a significant disturbance. If the process-to-process heat exchanger is not too large, this means that the furnace is operating under normal conditions. Then we are allowing more heat to go to utility at the separator cooler than we are generating in the reactor, and a bypass is not needed. For the HDA reactor with the quench loop, the furnace must be in operation at all times.

To ensure exothermic heat removal from the process, we are constrained by the process design to assign two control loops. We must control reactor inlet temperature with the furnace and control reactor exit temperature with the quench flow. Only by adjusting fuel to the furnace do we allow the reactor heat to be dissipated to the cooler. And only when the quench loop works do we guarantee that the furnace is operational. Because of this design we do not need a bypass line around the process-to-process heat exchanger.

Further discussion about energy management in the HDA process is given in Sec. 5.7.3.

Step 4. Hydrogen feed comes from a header and toluene feed is drawn from a supply tank. The benzene, methane, and diphenyl products go to headers or tanks. Hence we are not constrained to set production either via supply or demand. From the kinetic expressions given previously, we see that only three variables could be potentially dominant for the reactor: temperature, pressure, and toluene concentration. Pres-

sure is not a viable choice for production rate control because we want
to run at maximum system pressure and compressor capacity for yield
purposes. The other two are reactor inlet temperature and reactant
composition (since hydrogen is in excess, toluene is the limiting component). This gives us two viable options: change reactor inlet temperature or inlet toluene composition. Another effect of a change in production rate is that there has to be a net change of both the hydrogen and
toluene fresh feed rates into the process.

If we select temperature, we would like the reactor flow and composition to be nearly constant and we are constrained by the upper reactor
temperature limit of 1300°F. If we select toluene composition, we can
control it either directly or indirectly. If directly, a reactor feed composition analyzer is needed and is used to adjust either the fresh toluene
feed rate or the total reactor toluene feed rate. If indirectly, the separation section is used as an analyzer for toluene. This allows us to control
the total flow of toluene to the reactor (recycle plus fresh). Fresh toluene
feed flow is used to control toluene inventory reflected in the recycle
column overhead receiver level as an indication of the need for reactant
makeup. Controlling the total toluene flow sets the reactor composition
indirectly and is advantageous because it is less complicated and does
not require an on-line analyzer.

Step 5. The distillate stream from the product column is salable benzene. Benzene quality can be affected primarily by two components,
methane and toluene. Any methane that leaves in the bottoms of the
stabilizer column contaminates the benzene product. The easy separation in the stabilizer column allows us to prevent this by using a temperature to set column steam rate (boilup). Toluene in the overhead of
the product column also affects benzene quality. In this column the
separation between benzene and toluene is also fairly easy. As a result,
we can control product column boilup by using a tray temperature. To
achieve on-aim product quality control, we most likely would use an
on-line overhead composition analyzer to adjust the setpoint of this
temperature controller.

Important operational constraints have already been addressed with
the quench and reactor inlet temperature control loops set in Step 3.
If reactor inlet temperature is used to set production, we would include
an override on this controller to keep the reactor exit temperature
below the maximum allowable value. If toluene flow (fresh or recycle)
is used to set rate, then we would need an override on this to maintain
the minimum 5:1 ratio of hydrogen to aromatics in the reactor feed.

Step 6. The recycle flow of toluene should be fixed. We will soon see
what effect this has on the level control loops.

Four pressures must be controlled: in the three distillation columns and in the gas loop. In the stabilizer column, vapor product flow is the most direct manipulator to control pressure. In the product and recycle columns, pressure control can be achieved by manipulating cooling water flow to regulate overhead condensation rate. To maximize reactor yield, we open the separator overhead valve and run the compressor at maximum gas recycle rate to improve yield. We then have two choices to control gas loop pressure: purge or fresh hydrogen feed flow. Since pressure indicates hydrogen inventory in the gas recycle loop, we choose the hydrogen feed to control gas loop pressure.

Seven liquid levels are in the process: separator and two (base and overhead receiver) in each column. The most direct way to control separator level is with the liquid flow to the stabilizer column. Then stabilizer column overhead receiver level is controlled with cooling water flow and base level is controlled with bottoms flow. In the product column, distillate flow controls overhead receiver level and bottoms flow controls base level.

After these choices, we must now decide about level control in the recycle column. Contrary to the other columns, here the boilup ratio is large since the bottoms diphenyl flow is quite small compared with the toluene recycle rate. For this case, we choose to control base level with the steam flow because it has a much larger effect.

If we use distillate flow from the recycle column to control overhead receiver level, then we see that all of the flows around the liquid recycle loop are set on the basis of level. This violates our original statement to fix the toluene recycle flow. We are then left with the question how to control the overhead receiver level in the recycle column. We can use the fresh makeup toluene feed to control this level since it represents the toluene inventory in the process. Such a scheme limits large flowrate changes to the refining section and automatically ensures the component balance for toluene.

Step 7.   Methane is purged from the gas recycle loop to prevent it from accumulating, and its composition can be controlled with purge flow. Diphenyl is removed in the bottoms stream from the recycle column, where steam flow controls base level. Here we control composition (or temperature) with the bottoms flow. The inventory of benzene is accounted for via temperature and overhead receiver level control in the product column. Toluene inventory is accounted for via level control in the recycle column overhead receiver. Purge flow and gas-loop pressure control account for hydrogen inventory.

Table 10.7 summarizes the component balance control strategy.

Step 8.   We now can assign control loops within individual units. Cooling water flow to the cooler controls process temperature to the separator. Reflux to the stabilizer, product, and recycle columns can be flow-

TABLE 10.7 Component Material Balance

Component	Input	+Generation	−Output	−Consumption	=Accumulation Inventory controlled by
H_2	Fresh feed	$0.5V_R r_2$	Purge stream	$V_R r_1$	Pressure control of recycle gas loop
CH_4	0	$V_R r_1$	Purge stream	0	Composition control of recycle gas loop
C_6H_6	0	$V_R r_1$	Product stream	$2V_R r_2$	Temperature control in product column
C_7H_8	Fresh feed	0	0	$V_R r_1$	Level control in recycle column reflux drum
$C_{12}H_{10}$	0	$0.5V_R r_2$	Purge stream	0	Temperature control in recycle column

controlled because there is no requirement at the unit operations level to do anything beyond this. Step 9 discusses how the reflux flows should be set.

Step 9. The basic regulatory strategy has now been established (Fig. 10.2). We have some freedom to select several controller setpoints to optimize economics and plant performance. If reactor inlet temperature sets production rate, the setpoint of the total toluene flow controller can be selected to optimize reactor yield. However, there is an upper limit on this toluene flow to maintain at least a 5:1 hydrogen-to-aromatic ratio in the reactor feed since hydrogen recycle rate is maximized. The setpoint for the methane composition controller in the gas recycle loop must balance the trade-off between yield loss and reactor performance. Reflux flows to the stabilizer, product, and recycle columns must be determined on the basis of column energy requirements and potential yield losses of benzene (in the overhead of the stabilizer and recycle columns) and toluene (in the base of the recycle column). Since the separations are easy, in this system economics indicate that the reflux flows would probably be constant.

10.3 Dynamic Simulations

We have constructed a rigorous nonlinear dynamic model of the HDA process with TMODS. We have used the model to demonstrate that we have developed a workable control strategy for various disturbances, including changes in production rate. Other control strategies have

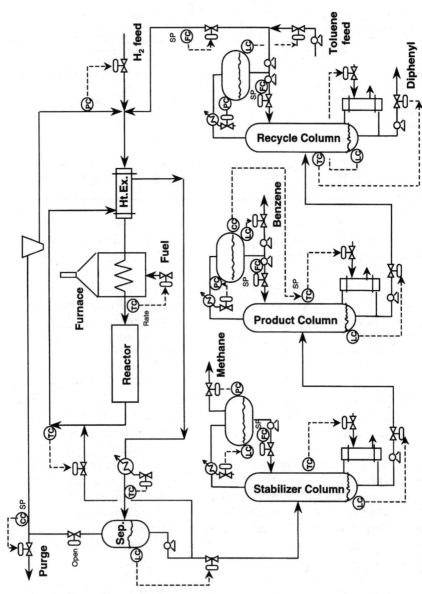

Figure 10.2 HDA process control strategy.

been previously proposed for this process that are significantly different than ours (Stephanopoulos, 1984, and Ponton and Laing, 1993). However, to our knowledge no simulation results or verifications have been reported.

Three simulation cases are discussed in this section. These cases differ in the size of the feed-effluent heat exchanger (FEHE) and in the use of heat-exchanger bypassing as part of the control structure used for reactor inlet temperature control. The effects of these design parameters were discussed in Chap. 5 in terms of the effectiveness of heat exchanger temperature control. Our purpose in this chapter is to demonstrate that the plantwide control structure developed by using the design procedure does indeed yield an effective, stable, base-level regulatory control system.

Two types of disturbances are used to test the response of the system: a step change in toluene recycle flowrate and a step change in the setpoint of the reactor inlet temperature controller. These two variables are the primary manipulators for production rate. In the results presented below we will explore which of the two is better. In addition we will see how several design parameters (FEHE area and heat-exchanger bypassing) impact the load response of the process.

We are particularly interested in two aspects of control structure effectiveness. First, how do these design parameters affect the control of product quality (benzene purity in the distillate stream from the product column)? Second, how do they affect the robustness of the control structure? We use *robustness* here to mean how large a step disturbance we can make and still have a stable response. The process is nonlinear, so performance will be a function of the magnitude of the forcing function. Almost all real processes and control structures can be made to fail if they are upset with a large enough disturbance. It is desirable, though, for a process and control structure to handle relatively large disturbances.

10.3.1   Control structure cases

Figure 10.2 gives the base-case plantwide control structure developed. Total toluene flowrate to the reaction section is flow-controlled. We will make step changes in this flow controller setpoint. Reactor inlet temperature is controlled by the firing rate in the furnace. No heat-exchanger bypass is shown in Fig. 10.2, but we will look at the effect of bypassing the FEHE. Control structure CS2 discussed in Chap. 5 adds a temperature control loop that controls furnace inlet temperature by manipulating the bypass flowrate around the FEHE. See Fig. 5.25.

Simulation results are discussed below. Stable control is achieved in all cases, but the use of a small FEHE and a large furnace gives a process that is the most robust to disturbances. As we will show, reactor inlet temperature setpoint changes of up to 22°F and toluene flowrate changes of up to 30 percent can be made while still maintaining stable response. The worst case is when a large heat exchanger and a small furnace are used. Toluene flowrate changes are limited to only 15 percent.

Another important result is that changing reactor inlet temperature disturbs the process less than changing toluene recycle flowrate. This means that reactor temperature is a better handle to use for changing production rate than is toluene recycle. This is consistent with the notion of using unit control instead of relying on variables controlled elsewhere in the process. In the results discussed below, we found the largest toluene recycle flowrate change that the system could handle. Then we found the change in the reactor temperature setpoint that gave approximately the same production rate change. This provides a fair comparison between the two alternative production rate manipulators. The system was stable for larger reactor inlet temperature changes than those reported, but these corresponded to larger production rate changes.

10.3.2 Heat-exchanger bypass (CS2 control structure) case

Figures 10.3 and 10.4 give simulation results for the case where control structure CS2 is used. This control scheme features heat-exchanger bypass to control furnace inlet temperature and furnace firing rate to control reactor inlet temperature. Remember that a large heat exchanger is used in this design. Several tests were run with increasingly larger magnitude disturbances until a limit was reached that still maintained a stable response. Toluene recycle flowrate can be changed 20 percent, which changes benzene production from about 270 to 210 lb · mol/h. The change in reactor inlet temperature that gives the same production rate change is about 18°F.

In Figure 10.3a the flowrates of the fresh feed streams of hydrogen and toluene are shown. An 18°F decrease in reactor inlet temperature is made at time equals 10 minutes, and then the temperature is returned to its normal value at time equals 125 minutes. The drop in temperature reduces reaction rates, so the flowrates of the fresh reactant feed streams are reduced. After a fairly short time lag, the benzene product rate also drops as shown in Figure 10.3b. The lower inlet temperature produces a lower reactor exit temperature, so less quench flow is required to maintain quench temperature (1150°F). Less heat-exchanger bypassing is required to maintain the furnace inlet temperature (1082°F) because the flowrate of the hot stream entering the FEHE has dropped.

(A)

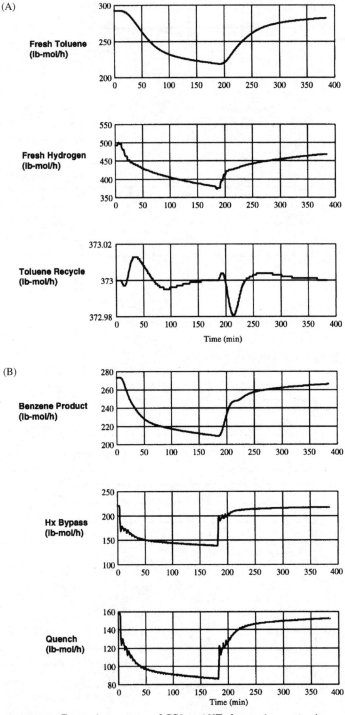

Figure 10.3   Dynamic response of CS2 to 18°F change in reactor inlet temperature.

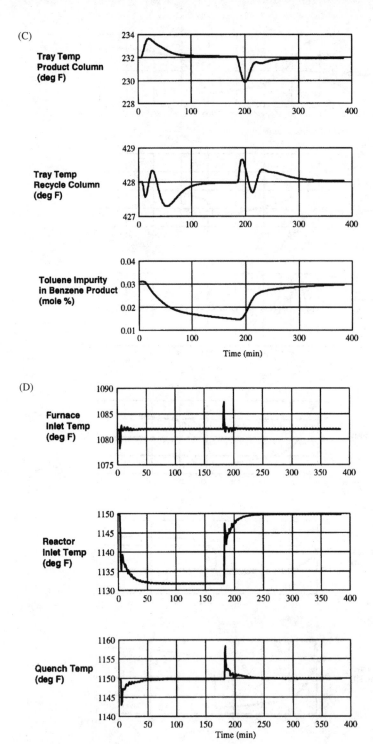

Figure 10.3 *(Continued)*

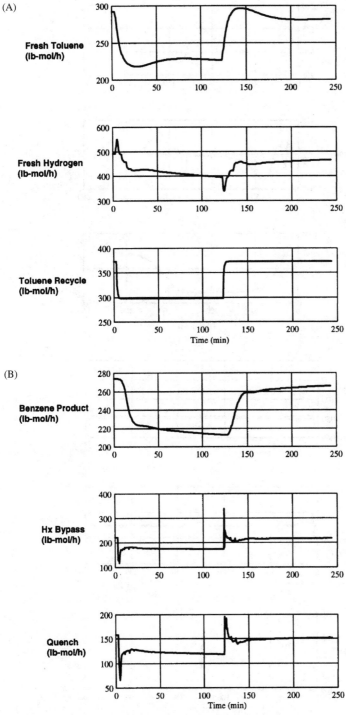

Figure 10.4 Dynamic response of CS2 to 20 percent change in toluene recycle flow.

(C)

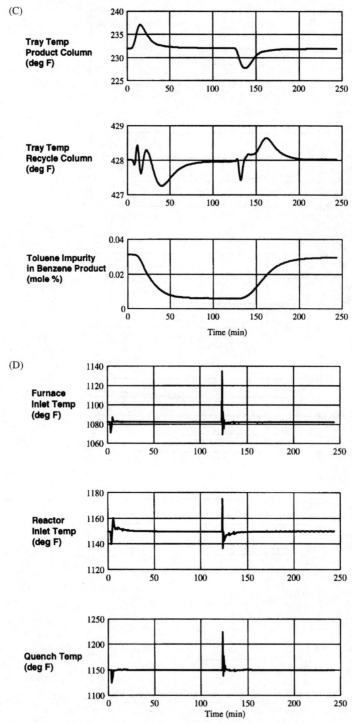

Figure 10.4 *(Continued)*

Figure 10.3c shows how the disturbance affects benzene product quality. The maximum deviation in the control tray temperature is about 2°F. Notice that a constant tray temperature does not give constant benzene purity, which changes by 0.015 mole %. Reflux flowrate is constant in the product column, so product purity improves as the feed rate to the column decreases.

Figure 10.4 gives results for 20 percent changes in the toluene recycle flowrate. The change in benzene production is the same as with the 18°F decrease in reactor inlet temperature. Now, however, the control tray temperature in the product column is disturbed about 5°F, compared to 2°F when reactor inlet temperature was changed.

Note that the changes in production rate occur more quickly when the toluene recycle flowrate handle is used, compared to the reactor inlet temperature handle. Fresh feed rates of toluene and hydrogen change more quickly, as does benzene product flowrate. So if rapid transitions in production rate are important, toluene recycle flowrate manipulation is better than reactor inlet temperature manipulation. If tight product quality control is more important, the opposite is true.

10.3.3 Large heat exchanger case

As shown in Figures 10.5 and 10.6, the largest toluene recycle flowrate change that can be handled with the large FEHE design is about 15 percent, compared to 20 percent when heat exchanger bypassing is used. This corresponds to a production rate change from 270 to 220 lb · mol/h, which is less than that achieved in the previous case. The corresponding reactor inlet temperature change is about 14°F. Reactor inlet temperature changes cause less upset to the process than toluene recycle flowrate changes. Reactor inlet temperature is not controlled as well as with the CS2 control strucuture.

10.3.4 Small heat exchanger case

Figures 10.7 and 10.8 give results for the small heat exchanger and large furnace case, which can tolerate the largest changes in productions rate. A 30 percent drop in toluene recycle flowrate is handled, giving a reduction in production from 270 to 180 lb · mol/h. The reactor inlet temperature change that gives about the same decrease in production rate is 22°F.

The response of the system to a reactor inlet temperature change is slower than that for a change in toluene recycle flowrate. Toluene recycle flowrate changes disturb the benzene product-quality control more than reactor inlet temperature changes.

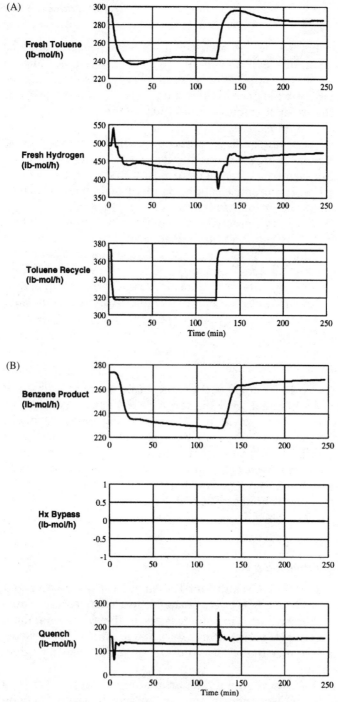

Figure 10.5 Dynamic response for large heat exchanger design with 15 percent change in toluene recycle flow.

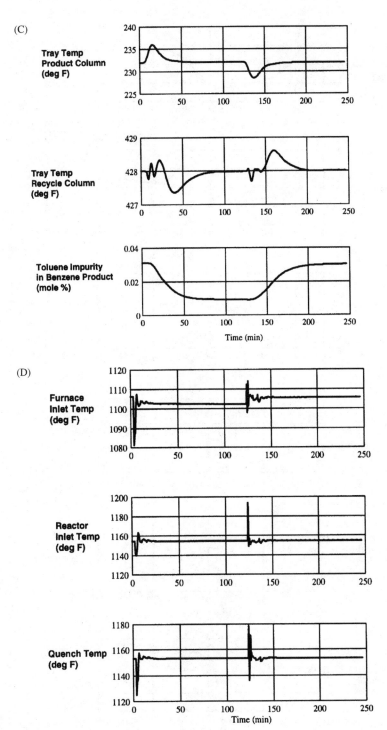

(C)

Tray Temp Product Column (deg F)

Tray Temp Recycle Column (deg F)

Toluene Impurity in Benzene Product (mole %)

Time (min)

(D)

Furnace Inlet Temp (deg F)

Reactor Inlet Temp (deg F)

Quench Temp (deg F)

Time (min)

Figure 10.5 *(Continued)*

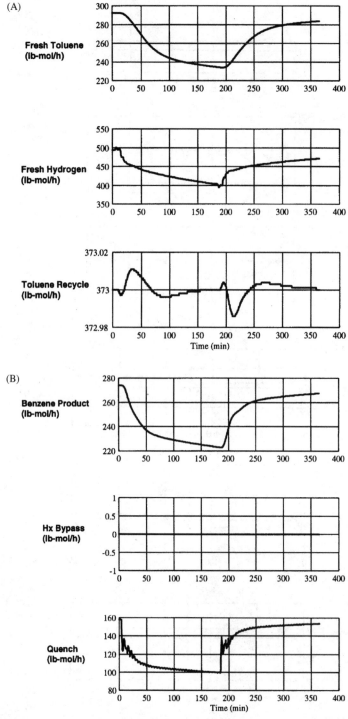

Figure 10.6 Dynamic response of large heat exchanger design with 14.4°F change in reactor inlet temperature.

(C)

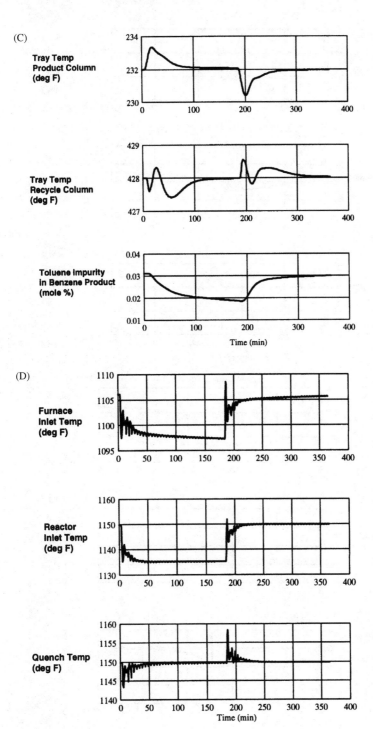

Figure 10.6 *(Continued)*

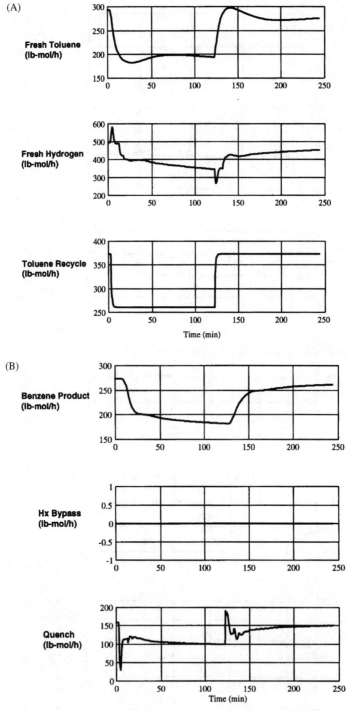

Figure 10.7 Dynamic response of small heat exchanger design with 30 percent change in toluene recycle flow.

(C)

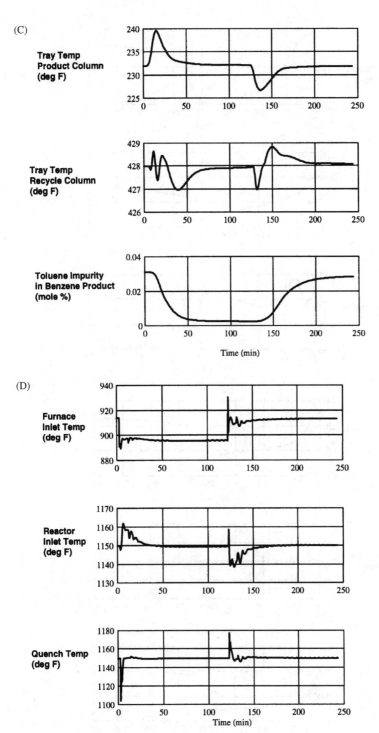

Figure 10.7 (Continued)

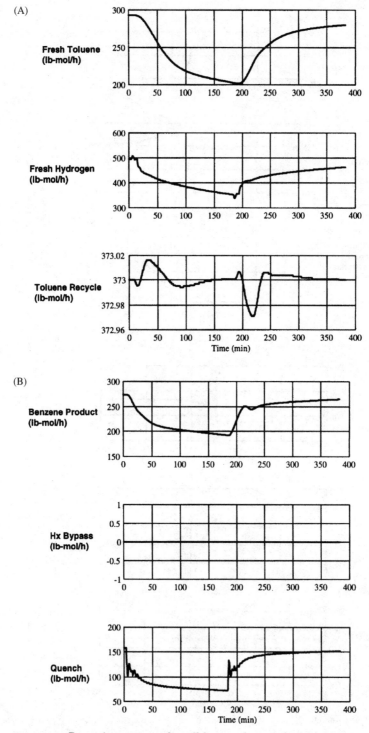

Figure 10.8 Dynamic response of small heat exchanger design with 21.6°F change in reactor inlet temperature.

(C)

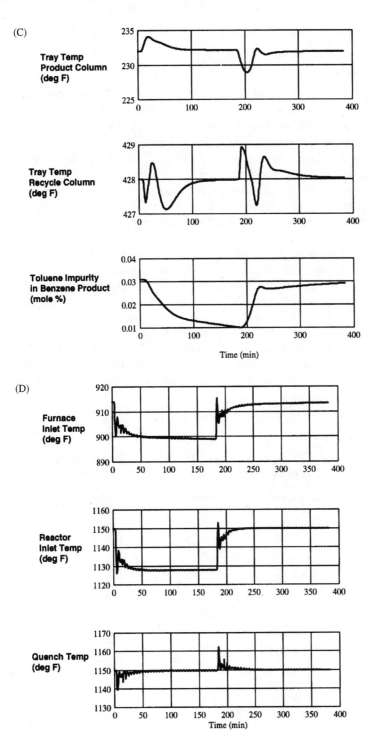

(D)

Figure 10.8 *(Continued)*

319

10.4   Conclusion

In this chapter we have applied the plantwide control design procedure to the HDA process. The HDA process is typical of many chemical process with many chemical components, many unit operations, several recycle streams, and energy integration. The steady-state design of the HDA process has been extensively studied in the literature, but no quantitative study of its dynamics and control has been reported.

An effective base-level regulatory control system has been developed and tested using a rigorous, nonlinear dynamic simulation of the entire system: tubular reactor, heat exchangers, and three distillation columns.

The impact of several design parameters has been explored. Control performance worsens when the steady-state economic optimum design, consisting of a large feed-effluent heat exchanger and a small furnace, is used. The most robust control is obtained when a small FEHE and a large furnace are employed.

Production rate changes using reactor inlet temperature as the manipulator cause less variability in benzene product quality in the HDA process. In addition, larger changes in production rate can be made by using reactor inlet temperature than can be achieved by making recycle toluene flowrate changes.

10.5   References

Douglas, J. M. *Conceptual Design of Chemical Processes*, New York: McGraw-Hill (1988).
Ponton, J. W., and Laing, D. M. "A Hierarchical Approach to the Design of Process Control Systems," *Chem. Eng. Res. Des.*, **71**, 181–188 (1993).
Stephanopoulos, G. *Chemical Process Control*, New York: Prentice-Hall (1984).
Zimmerman, C. C., and York, R. "Thermal Demethylation of Toluene," *I&EC Proc. Des. Dev.*, **3**, 254–258 (1964).

11

Vinyl Acetate Process

11.1 Introduction

The final example to illustrate our plantwide control design procedure comes from Luyben and Tyreus (1998), who present design details of an industrial process for the vapor-phase manufacture of vinyl acetate monomer. This process is uniquely suited for researchers pursuing process simulation, design, and control studies. It has common real chemical components in a realistically large process flowsheet with standard chemical unit operations, gas and liquid recycle streams, and energy integration.

The study was conveyed as if we had been assigned the task of designing the control system for a proposed new vinyl acetate process that is to be built. A particular preliminary design was given that had *not* been optimized. The data provided are what would typically be available or easily obtainable: (1) kinetic reaction parameters and physical property data, (2) a flowsheet structure with stream and equipment information, and (3) the location of control valves included in the preliminary design.

The industrial process for the vapor-phase manufacture of vinyl acetate monomer is quite common (Daniels, 1989) and utilizes widely available raw materials. Vinyl acetate is used chiefly as a monomer to make polyvinyl acetate and other copolymers. Hoechst-Celanese, Union Carbide, and Quantum Chemical are reported U.S. manufacturers. DuPont also currently operates a vinyl acetate process at its plant in LaPorte, Texas. To protect any proprietary DuPont information, all of the physical property and kinetic data, process flowsheet information, and modeling formulation in the published paper come from sources in the open literature. The process flowsheet is based upon the description in Report 15B by SRI International (1994). No relation, either implied or intended, exists between the published study and the DuPont process.

Figure 11.1 shows the eleven basic unit operations in the reaction section of the vinyl acetate process. Three raw materials, ethylene (C_2H_4), oxygen (O_2), and acetic acid (HAc), are converted into the vinyl acetate (VAc) product. Water (H_2O) and carbon dioxide (CO_2) are by-products. We assume that an inert component, ethane (C_2H_6), enters with the fresh ethylene feed stream. We consider the following two reactions:

$$C_2H_4 + CH_3COOH + \tfrac{1}{2} O_2 \rightarrow CH_2 = CHOCOCH_3 + H_2O \quad (11.1)$$

$$C_2H_4 + 3O_2 \rightarrow 2CO_2 + 2H_2O \quad (11.2)$$

The exothermic reactions occur in a reactor containing tubes packed with a precious metal catalyst on a silica support. Heat is removed from the reactor by generating steam on the shell side of the tubes. Water flows to the reactor from a steam drum, to which makeup water (boiler feeder water; BFW) is supplied. The steam leaves the drum as saturated vapor. The reactions are irreversible and the reaction rates have an Arrhenius-type dependence on temperature.

The following rate expressions were derived from the experimental kinetic data in Samanos et al. (1971) for a particular vinyl acetate catalyst.

$$r_1 = 0.1036 \exp\left(-3674/T\right) \frac{p_O\, p_E\, p_A\, (1+ 1.7p_W)}{[1 + 0.583p_O(1 + 1.7p_W)](1 + 6.8p_A)} \quad (11.3)$$

$$r_2 = 1.9365 \times 10^5 \exp\left(-10{,}116/T\right) \frac{p_O(1 + 0.68p_W)}{1 + 0.76p_O(1 + 0.68p_W)} \quad (11.4)$$

where r_1 has units of moles of vinyl acetate produced per minute per gram of catalyst and r_2 has units of moles of ethylene consumed per minute per gram of catalyst. T is the absolute temperature in kelvin and p_i is the partial pressure of component i (O is oxygen, E is ethylene, A is acetic acid, and W is water) in psia.

The ideal-gas standard state heat of reaction is -42.1 kcal/mol of vinyl acetate for r_1 and -316 kcal/mol of ethylene for r_2. These values are calculated from ideal-gas heats of formation from the DIPPR database. Thus the reactions are quite exothermic, particularly the combustion reaction to carbon dioxide, which also is more sensitive to temperature because of the higher activation energy.

The reactor effluent flows through a process-to-process heat ex-

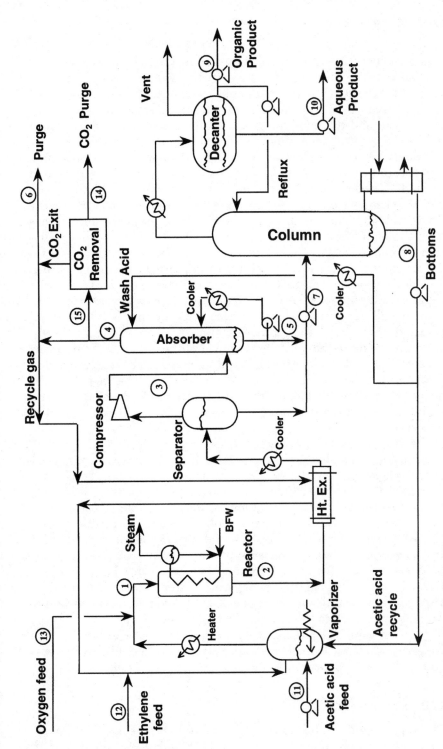

Figure 11.1 Vinyl acetate process flowsheet.

changer, where the cold stream is the gas recycle. The reactor effluent is then cooled with cooling water, and the vapor (oxygen, ethylene, carbon dioxide, and ethane) and liquid (vinyl acetate, water, and acetic acid) are separated. The vapor stream from the separator goes to the compressor and the liquid stream from the separator becomes a part of the feed to the azeotropic distillation column. The gas from the compressor enters the bottom of an absorber, where the remaining vinyl acetate is recovered. A liquid stream from the base is recirculated through a cooler and fed to the middle of the absorber. Liquid acetic acid that has been cooled is fed into the top of the absorber to provide the final scrubbing. The liquid bottoms product from the absorber combines with the liquid from the separator as the feed stream to the distillation column.

Part of the overhead gas exiting the absorber enters the carbon dioxide removal system. This could be one of several standard industrial CO_2 removal processes. Here we simplify this system by treating it as a component separator with a certain efficiency that is a function of rate and composition. The gas stream minus the carbon dioxide is split, with part going to the purge for removal of the ethane inert from the process. The rest combines with the large recycle gas stream and goes to the feed-effluent heat exchanger. The fresh ethylene feed stream is added. The gas recycle stream, the fresh acetic acid feed, and the recycle liquid acetic acid stream enter the vaporizer, where steam is used to vaporize the liquid. The gas stream from the vaporizer is further heated to the desired reactor inlet temperature in a trim heater using steam. Fresh oxygen is added to the gas stream from the vaporizer just prior to the reactor to keep the oxygen composition in the gas recycle loop outside the explosivity region.

The azeotropic distillation column separates the vinyl acetate and water from the unconverted acetic acid. The overhead product is condensed with cooling water and the liquid goes to a decanter, where the vinyl acetate and water phases separate. The organic and aqueous products are sent for further refining to another distillation section. Here we ignore the additional separation steps required to produce vinyl acetate of sufficient purity because there is no recycle from the refining train back to the reaction loop. The bottoms product from the distillation column contains acetic acid, which recycles back to the vaporizer along with fresh makeup acetic acid. Part of this bottoms stream is the wash acid used in the absorber after being cooled.

11.2 Process Data

The vapor-liquid equilibrium (VLE) data for the three nonideal component pairs are in Table 11.1. These data come from the *Vapor-Liquid*

TABLE 11.1   Wilson Parameters a_{ij} and Molar Volumes V_i

a_{ij}	VAc	H_2O	HAc	V_i, mL/mol
VAc	0	1384.6	−136.1	93.1
H_2O	2266.4	0	670.7	18.07
HAc	726.7	230.6	0	57.54

From DECHEMA *Vapor-Liquid Equilibrium Data Collection*, Vol. 1 (VAc–H_2O: Part 1b, p. 236; VAc–HAc: Part 5, p. 90; H_2O–HAc: Part 1, p. 127).

Equilibrium Data Collection in the Chemistry Data Series published by DECHEMA. VLE calculations are performed assuming an ideal vapor phase and a standard Wilson liquid activity coefficient model. This takes the form

$$\Lambda_{ij} = \frac{V_j}{V_i} \exp\left(-a_{ij}/RT\right) \qquad (11.5)$$

where T is the absolute temperature in K, R is the gas constant (1.987 cal/mol · K), and V_i is the molar volume of component i given in DECHEMA and listed in Table 11.1.

The Wilson parameters used for the VAc–H_2O pair are assumed to be the same as the parameters for ethyl acetate and water. The reason for this assumption is that no VLE data are presented in DECHEMA for vinyl acetate and water, but ethyl acetate and vinyl acetate are quite similar species and should behave essentially identically. The liquid-liquid equilibrium solubility data for the VAc–H_2O pair in the column decanter come from Smith (1942) extrapolated to the decanter temperature of 40°C.

Table 11.2 shows the pure component physical property data, which were obtained from the DIPPR database. These data include the molecular weight MW, the liquid specific gravity (relative to the density of

TABLE 11.2   Pure Component Physical Properties

Component	Molecular weight	Specific gravity	Latent heat, cal/mol	Liquid heat capacity $a - b$, cal/g · °C	Vapor heat capacity $a - b$, cal/g · °C
O_2	32	0.5	2,300	0.3–0	0.218–0.0001
CO_2	44.01	1.18	2,429	0.6–0	0.23–0
C_2H_4	28.05	0.57	1,260	0.6–0	0.37–0.0007
C_2H_6	30.05	0.57	1,260	0.6–0	0.37–0.0007
VAc	86.09	0.85	8,600	0.44–0.0011	0.29–0.0006
H_2O	18.02	1	10,684	0.99–0.0002	0.56–(−0.0016)
HAc	60.05	0.98	5,486	0.46–0.0012	0.52–0.0007

water), the latent heat of vaporization ΔH_v extrapolated to 0°C (in cal/mol), and the liquid c_p^l and vapor c_p^v heat capacity parameters. The heat capacity expressions have the following temperature dependence:

$$c_p = a + bt \tag{11.6}$$

where c_p is in cal/(g · °C) and t is the temperature in °C.

Component vapor pressures P^s in psia (Table 11.3) are calculated by using the Antoine equation, with the Antoine coefficients based on the DECHEMA volumes:

$$\ln P^s = A + \frac{B}{(t + C)} \tag{11.7}$$

where t is the temperature in °C. For the four gas components, the A parameters of the Antoine equation were estimated on the basis of the vapor pressure at the operating conditions in the absorber. The temperature dependence was removed to facilitate the dynamic simulation. However, in the case of ethylene and ethane, it was found that a small temperature dependence needed to be included for the bubble point calculations to function properly.

The process design assumes a production basis with new catalyst of 785 mol/min VAc and at the given conditions 85 mol/min CO_2 is also produced. For a plant with 90 percent operating utility, this corresponds to an annual production rate of 32×10^6 kg/yr, if the VAc rate is sustained over the life of the catalyst. The catalyst lifetime is assumed to be 1 year. Available on the plant are the following utilities: cooling tower water at a supply temperature of 30°C, steam at supply pressures of 50 and 200 psia, refrigeration at −25°C, and electricity and process water. Economic data for raw material and energy costs are listed in Table 11.4. Capital equipment and vessel cost data can be found in Guthrie (1969); these costs should be updated to current prices with the appropriate material of construction factors applied.

TABLE 11.3 Component Vapor Pressure Antoine Coefficients*

Component	A	B	C
O_2	9.2	0	273
CO_2	7.937	0	273
C_2H_4	9.497	−313	273
C_2H_6	9.497	−313	273
VAc	12.6564	−2984.45	226.66
H_2O	14.6394	−3984.92	233.426
HAc	14.5236	−4457.83	258.45

*$\ln P^s = A + B/(t + C)$, where P^s is in psia and t is in °C.

TABLE 11.4 Economic Data for Vinyl
Acetate Process

Item	Cost/unit
Acetic acid	$0.596/kg
Oxygen	$0.044/kg
Ethylene	$0.442/kg
Vinyl acetate	$0.971/kg
200-psia steam	$11/1000 kg
50-psia steam	$8.8/1000 kg
Cooling tower water	$0.02/1000 L
Process water	$0.15/1000 L
−25°C refrigeration	$0.12/h · ton
Electricity	$0.065/kWh

Tables 11.5 to 11.7 contain process stream data. These data come from the TMODS dynamic simulation and not from a commercial steady-state simulation package. The corresponding stream numbers are shown on the flowsheet in Fig. 11.1. Tables 11.8 to 11.10 list the process equipment and vessel data. In the simulation, all gas is removed in a component separator prior to the distillation column. This involves the liquid from the separator and the absorber. The gas is sent back and combines with the vapor product from the separator to form the vapor feed to the absorber. Figure 11.2a shows the temperature profile in the azeotropic distillation column.

The reactor is modeled in 10 sections in the axial direction. The reactor temperature profile is shown in Fig. 11.2b. The flowsheet design conditions are for a new catalyst with an activity of 1. However, the

TABLE 11.5 Process Stream Data, Part 1

	Reactor in	Reactor out	Absorber vapor in	Absorber vapor out	Absorber liquid out	Purge flow
Stream number	1	2	3	4	5	6
Flow, mol/min	19,250	18,850	16,240	15,790	1210	3
Temperature, °C	148.5	158.9	80	40.4	47.7	40.4
Pressure, psia	128	90*	128	128	128	128
O_2, mole fraction	0.075	0.049	0.057	0.058	0.001	0.059
CO_2	0.007	0.011	0.013	0.014	0.001	68†
C_2H_4	0.583	0.551	0.642	0.658	0.025	0.667
C_2H_6	0.216	0.221	0.256	0.263	0.010	0.266
VAc	0	0.043	0.021	0.002	0.255	0.002
H_2O	0.009	0.055	0.007	0.001	0.129	0.001
HAc	0.110	0.070	0.004	0.004	0.579	0.005

*Pressure drop in gas loop assumed to be in reactor.
†Moles per million.

TABLE 11.6 Process Stream Data, Part 2

	Column feed	Column bottoms	Organic product	Aqueous product	Fresh HAc feed
Stream number	7	8	9	10	11
Flow, mol/min	3820	2160	826	831	785
Temperature, °C	42.5	137.2	40	40	30
Pressure, psia	84	30	18	18	150
VAc, mole fraction	0.206	11*	0.950	0.002	0
H_2O	0.281	0.093	0.050	0.998	0
HAc	0.513	0.907	370*	370*	1

*Moles per million.

TABLE 11.7 Process Stream Data, Part 3

	Fresh C_2H_4 feed	Fresh O_2 feed	CO_2 purge	CO_2 removal in flow
Stream number	12	13	14	15
Flow, mol/min	831	521	85	6411
Temperature, °C	30	30	40.4	40.4
Pressure, psia	150	150	128	128
O_2, mole fraction	0	1	0	Same
CO_2	0	0	1	as
C_2H_4	0.999	0	0	stream
C_2H_6	0.001	0	0	4

TABLE 11.8 Reactor and Vaporizer Equipment Data

Catalyst weight	2590 kg
Catalyst porosity	0.8
Catalyst bulk density	0.385 kg/L
Catalyst heat capacity	0.23 cal/g · °C
Overall heat transfer coefficient	150 kcal/h · °C · m²
Number of tubes	622
Tube length	10 m
Tube diameter	3.7 cm
Circumferential heat transfer area	725 m²
Shell side temperature	133 °C
Reactor heat duty	2.8×10^6 kcal/h
Steam drum volume	2 m³
BFW to steam drum	79.5 kg/min
Reactor feed heater duty	5.3×10^5 kcal/h
Vaporizer duty	1.3×10^6 kcal/h
Vaporizer total volume	17 m³
Vaporizer working level volume	4 m³
Vaporizer temperature	119 °C

TABLE 11.9   FEHE, Separator, and Absorber Equipment Data

FEHE duty	4.4×10^5 kcal/h
FEHE hot outlet temperature	134 °C
FEHE UA	6800 kcal/h $\cdot$ °C
Separator:	
Cooler duty	2.7×10^6 kcal/h
Volume	15 m³
Working level volume	8 m³
Gas loop volume	170 m³
Approximate compressor size	350 kW
Absorber:	
Base volume	8 m³
Bottom section	2 theoretical stages
Top section	6 theoretical stages
Stage efficiency	50 %
Tray holdup	14 kmoles
Liquid recirculation	15 kmol/min
Cooler duty	6.5×10^5 kcal/h
Wash acid feed	756 mol/min
Wash acid cooler duty	1.3×10^5 kcal/h

catalyst does deactivate over the course of operation. This deactivation via sintering is a nonlinear function of operating time (t_{yr}) and temperature, since higher temperatures within the tubes (t_{tube}) promote deactivation. The activity a decays exponentially with time from 1.0 to 0.8 after 1 year according to

$$a = f(t_{tube}) \exp(-t_{yr}/0.621) \tag{11.8}$$

If the tube temperature has not exceeded 180°C, then $f(t_{tube}) = 1$. Above this temperature, then $f(t_{tube}) = \exp[-(t_{tube} - 180)/50]$, where t_{tube} is in °C.

Catalyst selectivity (SEL) determines the fraction of the ethylene consumed that makes the desired vinyl acetate product.

$$SEL = 100 \frac{\text{mol/min VAc}}{\text{mol/min VAc} + 0.5 \text{ mol/min CO}_2} \tag{11.9}$$

For our new catalyst conditions, the selectivity is 94.84 percent. At a

TABLE 11.10   Column and Decanter Equipment Data

Theoretical stages	20
Feed stage	15 from bottom
Stage efficiency	50 %
Tray holdup	2.3 kmol
Reboiler duty	4.0×10^6 kcal/h
Condenser duty	3.9×10^6 kcal/h
Base working level volume	6 m³
Decanter working level volume	5 m³

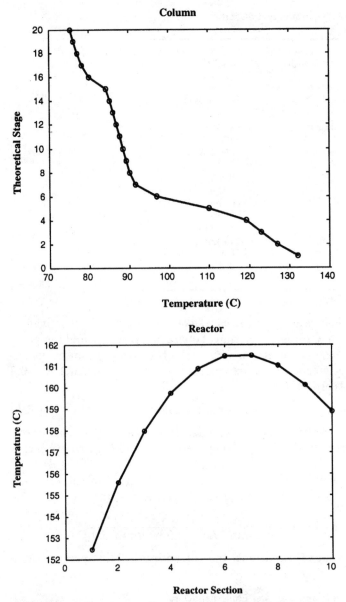

Figure 11.2 Temperature profiles. (*a*) Azeotropic distillation column;
(*b*) reactor.

catalyst activity of 0.8, higher reactor temperatures are required to
achieve about the same VAc production rate, increasing the production
rate of CO_2 to 126 mol/min and reducing the selectivity to 92.4 percent.

   The CO_2 removal system is assumed to be a component separator
that removes just carbon dioxide at a certain efficiency, which is the

fraction in the feed leaving in the CO_2 purge. This efficiency (Eff) is a function of the feed rate (F_{CO_2} in mol/min) and composition (x_{CO_2} in mole fraction). At the design conditions, the efficiency is 0.995 for a feed rate of 6410 mol/min at 0.014 mole fraction CO_2. The maximum allowable feed rate to the CO_2 removal system is 8000 mol/min set by its capacity. The following correlation determines the system efficiency:

$$\text{Eff} = 0.995 - 3.14 \times 10^{-6}(F_{CO_2} - 6410) - 32.5(x_{CO_2} - 0.014) \quad (11.10)$$

where the efficiency must lie between 0 and 1.

11.3 Plantwide Control Strategy

Step 1. For this process we must be able to set the production rate of vinyl acetate while minimizing yield losses to carbon dioxide. During the lifetime of the catalyst charge, catalyst activity decreases and the control system must operate under these different conditions. To maintain safe operating conditions, the oxygen concentration in the gas loop must remain outside the explosivity region for ethylene. The azeotropic distillation column must produce an overhead product with essentially no acetic acid and a bottoms product with no vinyl acetate. The absorber must recover essentially all of the vinyl acetate, water, and acetic acid from the gas recycle loop to prevent yield losses in the CO_2 removal system and purge.

Step 2. There are 26 control degrees of freedom in this process. They include: three feed valves for oxygen, ethylene, and acetic acid; vaporizer and heater steam valves; reactor steam drum liquid makeup and exit vapor valves; vaporizer overhead valve; two coolers and absorber cooling water valves; separator base and overhead valves; absorber overhead, base, wash acid, and liquid recirculation valves; gas valve to CO_2 removal system; gas purge valve; distillation column steam and cooling water valves; column base, reflux, and vent valves; and decanter organic and aqueous product valves.

Step 3. Energy management is critically important because of the highly exothermic reactions and potential for runaway or catalyst damage at high temperatures. By design, heat is removed from the reactor via transfer from the tubes to the shell, generating steam. Hence reactor temperature is controlled by steam temperature, which is set by controlling the pressure in the steam drum via the steam exit valve. This is a good example of how a degree of freedom needs to be used based upon the process design. If temperature measurements are available along the length of the tubes, then a reactor peak temperature or a profile can be controlled. Otherwise, the controlled variable is reactor exit temperature.

The reactor effluent stream is cooled in a process-to-process heat exchanger with the gas recycle stream. A bypass line and control valve are necessary here only if we want to control one of the exchanger exit temperatures. If this exchanger is designed for only vapor flow, then the hot-side exit temperature must be controlled to a value above the dewpoint temperature by manipulating the bypass valve around the exchanger on the cold side (to avoid a control valve on hot stream side). The bypass line would have to be added if it were not included in the original design or the heat exchanger must be redesigned to handle two-phase flow.

Step 4. Ethylene and oxygen makeup feeds come from headers and the acetic acid feed is drawn from a supply tank. The vinyl acetate and water products go to downstream units. As a result, there are no design constraints that require production rate to be set either on supply or demand. Therefore, we look at reactor conditions to determine how to change production rate. Because the reactor feed contains both excess ethylene and acetic acid, manipulating the partial pressure of either component would not be effective. The partial pressure of oxygen is constrained by the safety limit, and once this is reached no further adjustments could be made. Pressure is limited by the process equipment design maximum. Hence the most direct handle for setting production rate is by changing the reactor exit or peak temperature.

Alternatively we could use the fresh oxygen feed flow to set production rate since it is the limiting component. However, there are two issues with this choice. Since oxygen is not completely consumed, we must worry about its accumulation in the system (component balance), which is constrained here by the safety limit. If oxygen were completely consumed, we must still worry about the reactor inlet oxygen composition because of the safety constraint. In either case, with oxygen feed rate we would have to control oxygen composition with some other variable to change production rate safely.

Step 5. The azeotropic distillation column does not produce the final salable vinyl acetate product. Its primary role is to recover and recycle unreacted acetic acid and to remove from the process all of the vinyl acetate and water produced. So we want little acetic acid in the overhead because this represents a yield loss. Also, the bottoms stream should contain no vinyl acetate since it polymerizes and fouls the heat-exchange equipment at the elevated temperatures of the column base and the vaporizer. Hence we have two control objectives: base vinyl acetate and top acetic acid compositions. And we have two manipulators, steam and organic reflux flows. The temperature profile (Fig. 11.2a) has a sharp break representing the change in vinyl acetate composition near the bottom of the column. Column steam (boilup) is the appropriate choice for temperature control because of its fast response compared with reflux. Then the overhead acetic acid composition must be controlled with reflux.

The overriding safety constraint in this process involves oxygen concentration in the gas loop, which must remain below 8 mole % to remain outside the explosivity envelope for ethylene mixtures at process conditions. The most direct manipulated variable to control oxygen composition at the reactor feed is the fresh oxygen feed flow.

If in Step 4 we had chosen to set production rate by flow-controlling the fresh oxygen feed, then we would need an alternative manipulator to control oxygen composition. The only choice would be to use reactor temperature. We now can consider which choice, safety or production rate, is better. Temperature is *not* the most direct handle to control oxygen composition since its effectiveness hinges on incomplete oxygen conversion in the reactor and oxygen recycle. Because of the safety implications, we would choose to use fresh oxygen flow to control reactor inlet oxygen composition, which means production rate is set via reactor temperature.

Step 6. Two pressures must be controlled: in the column and in the gas loop. The most direct handle to control column pressure is by manipulating the vent stream from the decanter. We have three choices to control gas loop pressure: purge flow, flow to the CO_2 removal system, and the fresh ethylene feed flow since fresh oxygen flow has been previously selected. Both the purge flow and the flow to the CO_2 removal system are small relative to the gas recycle flowrate. Any changes in either one would not have a large effect on gas loop pressure. Since ethylene composes a substantial part of the gas recycle stream, pressure is a good indication of the ethylene inventory. So we choose the fresh ethylene feed flow to control gas recycle loop pressure.

Note that there are three vessels (vaporizer, separator, absorber) within the gas loop where apparently pressure can also be controlled. However, these pressures actually cannot be selected arbitrarily once the compressor capacity and gas loop pressure drop are established. In fact, to minimize pressure drop in the gas loop we would open completely or remove the overhead vapor control valves on these units, saving both compressor and valve costs.

Seven liquid levels are in the process: vaporizer, reactor steam drum, separator, absorber, column base, and two decanter layers. Control of the decanter levels is straightforward. The organic product flow controls the organic phase inventory; the aqueous product flow controls the aqueous phase inventory. Reactor steam drum level is maintained with boiler feed water makeup flow.

The most direct way to control the remaining levels would be with the exit valves from the vessels. However, if we do this we see that all of the flows around the liquid recycle loop would be set on the basis of levels, which would lead to undesirable propagation of disturbances. Instead we should control a flow somewhere in this loop. Acetic acid is the main component in the liquid recycle loop. Recycle and fresh acetic acid feed determine the component's composition in the reactor feed. A reasonable choice at this point is to control the total acetic acid feed stream flow into the vaporizer. This means that we can use the fresh acetic acid feed stream to control column base level, since this is an indication of the acetic acid inventory in the process. Vaporizer level is then controlled with the vaporizer steam flow and separator and absorber levels can be controlled with the liquid exit valves from the units.

Step 7.   Ethane is an inert component that enters with the ethylene feed. It can be removed from the process only via the gas purge stream, so purge flow is used to control ethane composition. Carbon dioxide is an unwanted by-product that leaves in the CO_2 removal system. As long as the amount of carbon dioxide removed is proportional in some way to the CO_2 removal system feed, we can use this valve to control carbon dioxide composition. Oxygen inventory is accounted for via composition control with fresh oxygen feed. Inventory of ethylene can be controlled to maintain gas loop pressure, since ethylene composes the bulk of the gas recycle.

Acetic acid inventory is regulated by using the fresh acetic acid feed to control base level in the distillation column. The temperature control loop in the distillation column achieves vinyl acetate composition control. Water, however, is an intermediate component with a boiling point between vinyl acetate and acetic acid. The inventory of water in the process will not be automatically accounted for by controlling those other two components. Instead we must use reflux flow to control the water composition in the bottoms stream. Otherwise, there is no regulation of water to ensure it is removed from the process. By using column reflux to control base water, we are forced to give up on using reflux to control acetic acid recovery (Step 5). To ensure that the acetic acid composition in the decanter is acceptable, the column must have a sufficient number of stages. This means we may have to revisit its design.

Table 11.11 summarizes the component balance control strategy.

TABLE 11.11   Component Material Balance

Component	Input	+Generation	−Output	−Consumption	=Accumulation Inventory controlled by
O_2	Fresh feed	0	0	$V_R(0.5r_1 + 3r_2)$	Composition control of reactor feed
CO_2	0	$2V_R r_2$	CO_2 removal	0	Composition control of recycle gas loop
C_2H_4	Fresh feed	0	0	$V_R(r_1 + r_2)$	Pressure control of recycle gas loop
C_2H_6	Fresh feed	0	Purge stream	0	Composition control of recycle gas loop
VAc	0	$V_R r_1$	Product stream	0	Temperature control in distillation column
H_2O	0	$V_R(r_1 + 2r_2)$	Product stream	0	Column base composition control with reflux
HAc	Fresh feed	0	0	$V_R r_1$	Level control in column base

Step 8. Several control valves now remain unassigned. Steam flow to the trim heater controls reactor inlet temperature. Cooling water flow to the trim cooler is used to control the exit process temperature and provide the required condensation in the reactor effluent stream. Liquid recirculation in the absorber is flow-controlled to achieve product recovery, while the cooling water flow to the absorber cooler controls the recirculating liquid temperature. Acetic acid flow to the top of the absorber is flow-controlled to meet recovery specifications on the overhead gas stream. Cooling water flow to the cooler on this acetic acid feed to the absorber is regulated to control the stream temperature. Cooling water flow in the column condenser controls decanter temperature.

Step 9. We have now established the basic regulatory plantwide control strategy (Fig. 11.3). Based upon the heuristic established by Fisher et al. (1988) that recycle gas flows should be maximized to improve reactor yield, we open or remove the separator, vaporizer, and absorber overhead valves and run the compressor full out. To minimize the decanter temperature for improved organic recovery, the column condenser cooling water is set at maximum flow. Optimization of several controller

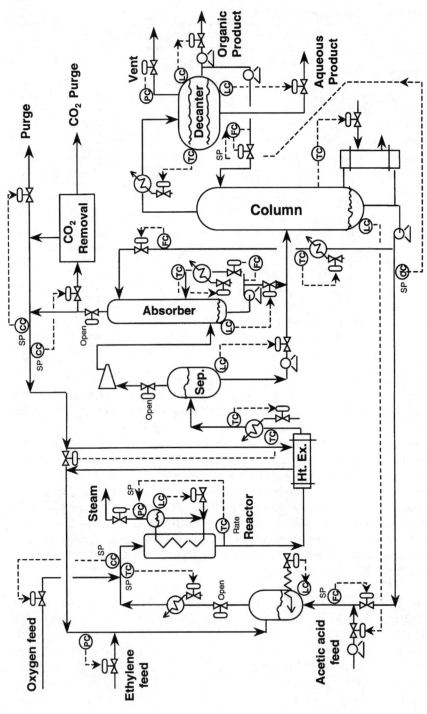

Figure 11.3 Vinyl acetate process control strategy.

setpoints can be based upon economics. We must balance the trade-offs in maximizing vinyl acetate production and recovery with minimizing carbon dioxide production and energy consumption. This involves considering the reactor temperature setpoint, reactor feed temperature setpoint, composition of carbon dioxide and ethane in the gas recycle loop, oxygen composition setpoint (up to the maximum constraint), total flowrate of acetic acid to the vaporizer, and water composition in the recycle acetic acid. Additionally, economic evaluations must account for the effects of catalyst deactivation with time over the duration of the charge. To achieve the same vinyl acetate production rate, we must increase reactor temperature. However, this produces more carbon dioxide representing yield loss. Raw material costs, energy costs, and product price will all affect plant operation.

11.4 Dynamic Simulations

We have constructed a rigorous nonlinear first-principles dynamic model of this process with TMODS. We have used the model to test the control strategy and show that it does provides effective control of the vinyl acetate monomer process.

Figures 11.4 to 11.9 present some results of the rigorous dynamic simulation to various disturbances. Because of the model size, many different variables could be plotted, but we have tried to include the key ones. Some of the dynamic behavior turns out to be not intuitively obvious. But the most important comment to make at the start is these results demonstrate that the control scheme developed with our design procedure works! We have generated a simple, easily understood regulatory control strategy for this complex chemical process that holds the system at the desired operating conditions.

The critical product-quality and safety-constraint loops were tuned by using a relay-feedback test to determine ultimate gains and periods. The Tyreus-Luyben PI controller tuning constants were then implemented. Table 11.12 summarizes transmitter and valve spans and gives controller tuning constants for the important loops. Proportional control was used for all liquid levels and pressure loops.

11.4.1 Changes in reactor temperature

Figure 11.4 shows what happens when the production rate handle (reactor exit temperature) is changed. The starting conditions are the base-case design where reactor exit temperature is 159°C. The reactor temperature controller is tuned at this operating point. Step changes of 8°C at time 5 minutes and 120 minutes are made in the setpoint of the reactor temperature controller (Figure 11.4a). Decreasing the

TABLE 11.12   Controller Settings for Certain Loops

Loop	Base case K_c	Base case τ_I, min	High-temperature case K_c	High-temperature case τ_I, min
$T_R - P_S^{set}$	10	5.5	0.6	11
$P_S - F_S$	20	1	20	1
$x_{B,W} - R$	0.2	60	0.2	60

P_S transmitter span: 40–60 psia
T_R transmitter span: 0–200°C
$x_{B,W}$ transmitter span: 0–10 wt %
R transmitter span: 0–7600 mol/min
T_R is reactor temperature, P_S is steam drum pressure, F_S is steam flowrate
$x_{B,W}$ is column base water composition, R is column reflux flow.

temperature reduces the rate of production of vinyl acetate but increases selectivity, i.e., the fraction of ethylene consumed producing vinyl acetate. The fresh feeds of oxygen and ethylene decrease fairly quickly (30 minutes). The fresh feed of acetic acid changes much more slowly. The changes in the reactor and the gas recycle are fast because the process has a gas-phase reactor. The changes in the column and liquid recycle are much slower because of the large liquid residence times. Note that the reactor production of VAc shown in Fig. 11.4b is the instantaneous reaction rate in the reactor, not the flowrate of organic product leaving the decanter. The flowrate of this stream changes quite slowly due to the large holdup in the decanter. Oxygen composition remains within about 0.1 mol % of setpoint. The step increase in reactor temperature is slightly more oscillatory than the decrease, but overall the performance is quite satisfactory for a change of this size.

Figure 11.5 gives results for a step increase in the reactor temperature controller setpoint from 159 to 165°C. It is clear that the loop becomes more oscillatory at the higher temperature. The loop becomes unstable if the setpoint is raised above 170°C. The reason for this change in the closed-loop damping coefficient is the change in reactor dynamics. This reactor is highly nonlinear, so we would not expect a linear controller to be effective over a wide range of operating conditions. If the reactor temperature controller is retuned at a higher 180°C operating point, we find (as shown in Table 11.12) that the controller gain is much smaller (0.6 versus 10 at the lower temperature). This is due to the change in the difference between the reactor exit temperature and the coolant temperature as temperature increases. Section 4.5.3 discusses this change in the reactor process gain with temperature. The higher temperature operating point also requires that the controller reset time be about twice that of the lower temperature operating point (larger ultimate period, slower closed-loop response).

(A)

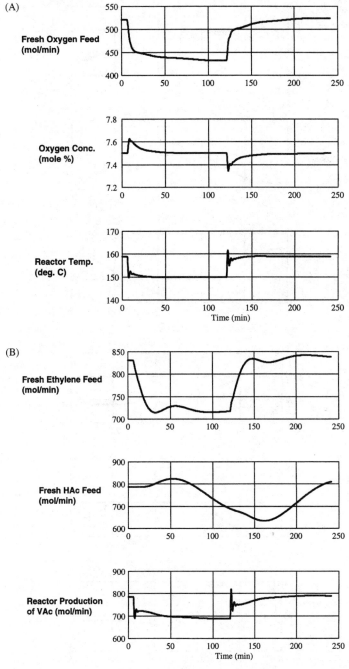

(B)

Figure 11.4 Dynamic response of base case to 8°C decrease in reactor temperature.

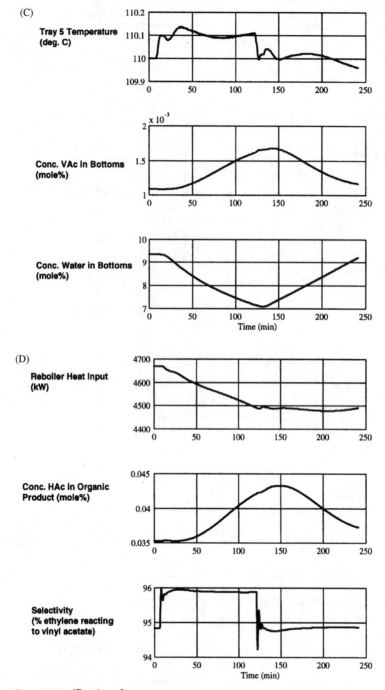

Figure 11.4   *(Continued)*

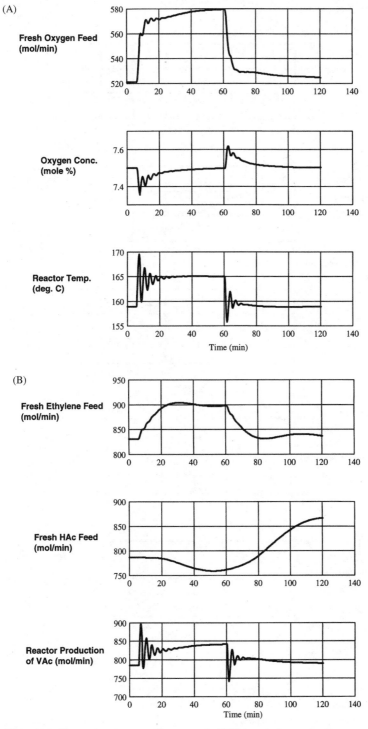

Figure 11.5 Dynamic response of base case to 6°C increase in reactor temperature.

(C)

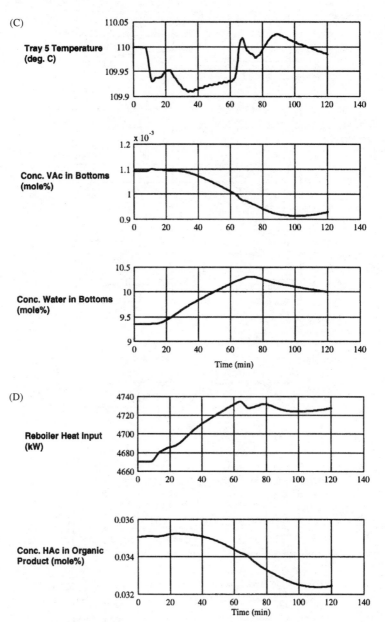

Tray 5 Temperature (deg. C)

Conc. VAc in Bottoms (mole%)

Conc. Water in Bottoms (mole%)

(D)

Reboiler Heat Input (kW)

Conc. HAc in Organic Product (mole%)

Figure 11.5 *(Continued)*

Figure 11.6 shows how the process responds to changes in reactor temperature setpoint when the initial condition is at the higher temperature steady state and the temperature controller has been retuned. The temperature control loop becomes significantly more sluggish as we decrease the temperature compared with the performance at the lower temperature steady state. Clearly a nonlinear controller (using gain scheduling) would be beneficial if operation at different steady states is required.

Increasing the reactor temperature setpoint increases the production rate of vinyl acetate, so there must ultimately be net increases in all three fresh reactant feed streams. Oxygen and ethylene flows respond fairly quickly within about 20 minutes. However, the acetic acid feed actually decreases for the first 60 minutes in response to an increase in column base level. These results demonstrate the slow dynamics of the liquid recycle loop and illustrate the need for controlling the total acetic acid flow to the reactor so that the separation section does not see these large swings in load ("snowball effect"). The variability is absorbed by the fresh feed makeup stream.

11.4.2 Loss of column feed pumps

Figure 11.7 gives results for a 5-minute shutdown of the distillation column feed pump. The flowrate of bottoms from the column drops quickly for about 20 minutes. The column tray temperature controller recovers in about 1 hour. Distillate and bottoms compositions are still changing after 5 hours.

The drop in bottoms flow requires a large increase in the flowrate of fresh acetic acid makeup so that the total acetic acid flowrate stays constant. Fresh acetic acid increases from about 750 up to 2200 mol/h in 30 minutes. This illustrates that the pumps and control valves used in a fresh feed makeup system must be designed to provide much wider rangeability than we would predict from just steady-state conditions alone.

Note that the reactor is affected, even with a constant total acetic acid flow. Both the ethylene and the oxygen fresh feeds increase temporarily, indicating an increase (though not permanent) in reaction rate. This is due to complicated interactions within the process dealing with vaporizer performance, fresh acetic acid feed and bottoms temperatures, and the reactant concentrations as they affect reactor performance.

11.4.3 Change in acetic acid recycle flowrate

Figure 11.8 gives the dynamic responses of the process for a 20 percent increase in the flowrate of total acetic acid to the vaporizer. The flow is

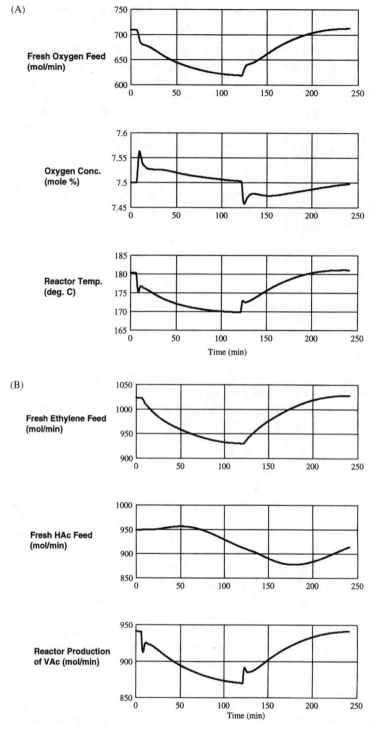

Figure 11.6 Dynamic response of high temperature case to 10°C decrease in reactor temperature.

(C)

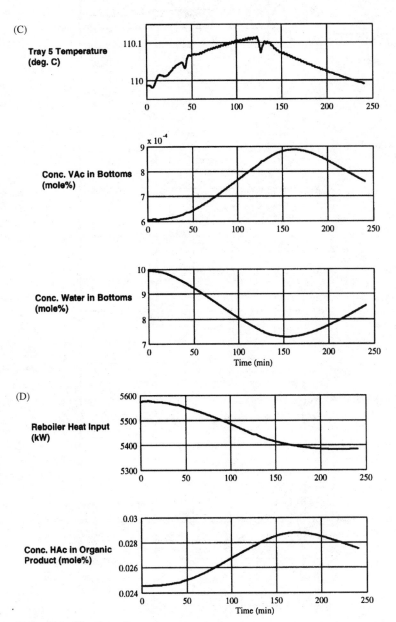

Figure 11.6 *(Continued)*

(A)

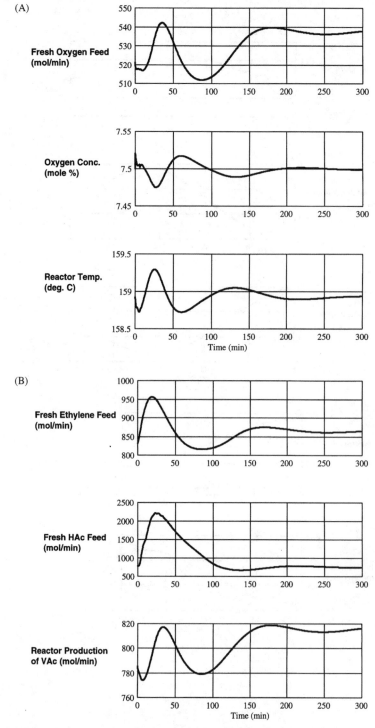

Figure 11.7 Dynamic response of base case to 5-minute shutoff of column feed.

(C)

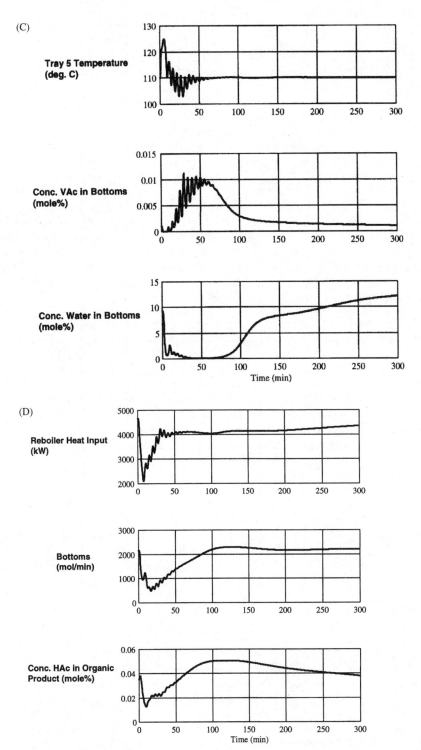

Tray 5 Temperature (deg. C)

Conc. VAc in Bottoms (mole%)

Conc. Water in Bottoms (mole%)

Time (min)

(D)

Reboiler Heat Input (kW)

Bottoms (mol/min)

Conc. HAc in Organic Product (mole%)

Time (min)

Figure 11.7 *(Continued)*

347

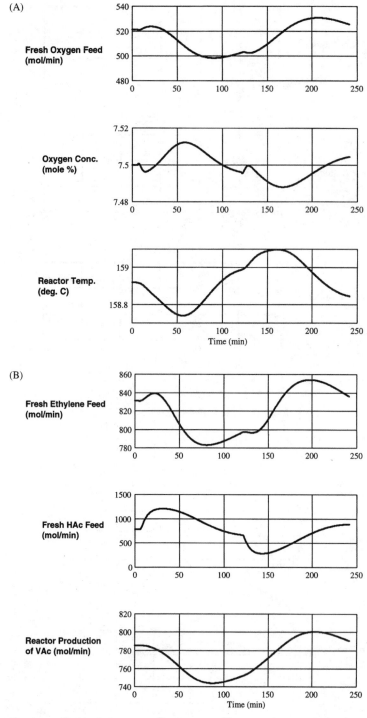

Figure 11.8 Dynamic response of base case to 20 percent increase in acetic acid recycle flow.

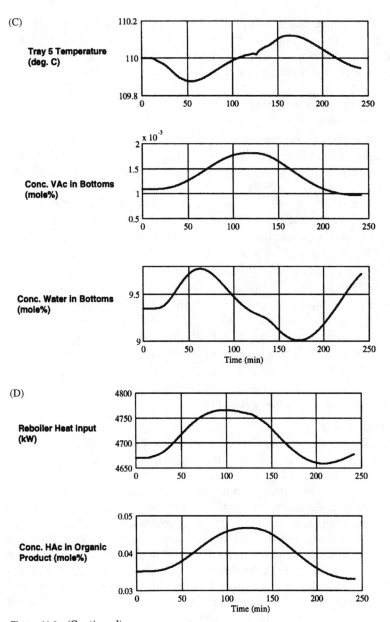

(C)

Tray 5 Temperature (deg. C)

Conc. VAc in Bottoms (mole%)

Conc. Water in Bottoms (mole%)

(D)

Reboiler Heat Input (kW)

Conc. HAc in Organic Product (mole%)

Figure 11.8   *(Continued)*

returned back to its initial condition after 2 hours. Note that the increase in acetic acid (one of the reactants) results in a decrease in the reaction rate (fresh feed flowrates of both oxygen and ethylene decrease). This is due to the unusual reaction kinetics discussed in Sec. 11.1.

11.4.4 Change in column base water composition

Figure 11.9 shows the effects of raising the setpoint of the bottoms water composition controller from 9 to 18 mole %. Because this loop is very slow (Table 11.12 shows a 60-minute reset time when a 3-minute deadtime is used with a 3-minute sampling time), conditions are still changing after 6 hours. Reactor temperature and oxygen concentration are very tightly controlled, but there are slow increases in the flowrates of fresh oxygen and ethylene. Ethylene fresh feed actually decreases for the very long period of time.

11.4.5 Summary

We could go on for many pages about the many interesting dynamic and steady-state effects observed in this complex nonlinear process. The primary message, however, is that a base-level regulatory control system has been designed that provides effective stable control of this complex process.

11.5 On-Demand Control Structure

The control system shown in Fig. 11.3 was developed under the assumption that we were free to select the production-rate handle. We decided to establish throughput indirectly by adjusting the setpoint of the reactor temperature controller.

In this section we illustrate how the control scheme is modified if a different control criterion is specified. Suppose business objectives dictate that the organic product from the decanter must be an "on-demand" stream, i.e., a downstream unit or customer sets the desired flowrate of this stream and the plant must immediately supply the requested flowrate of organic product. In this situation the organic product flowrate will be flow controlled, with the setpoint of the flow controller coming from the downstream consumer. A similar case was considered in Chap. 8 with the Eastman process.

Fig. 11.10 shows a modified control structure that incorporates this new on-demand criterion. It differs from the control scheme shown in Fig. 11.3 in only three loops:

1. The flowrate of the organic product from the decanter is flow controlled.

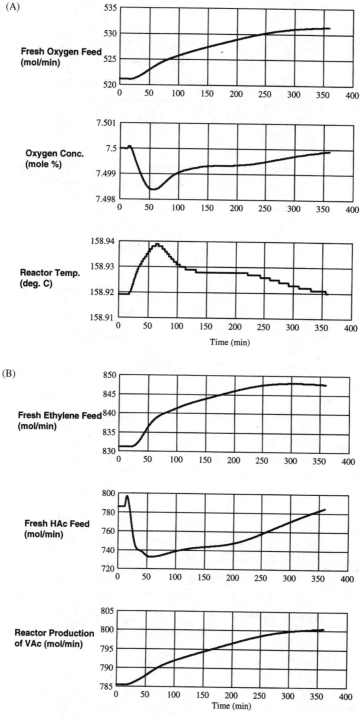

Figure 11.9 Dynamic response of base case to increase in column base water composition from 9 to 18 mol %.

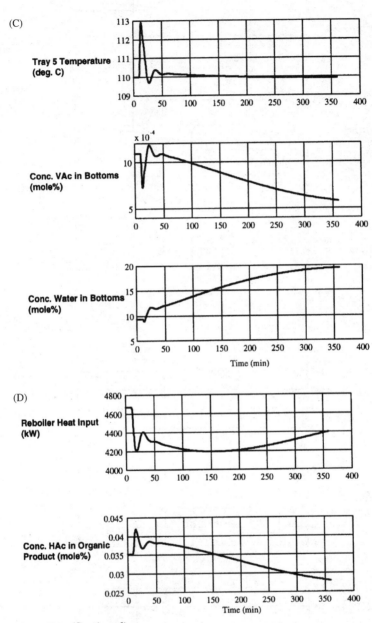

Figure 11.9 *(Continued)*

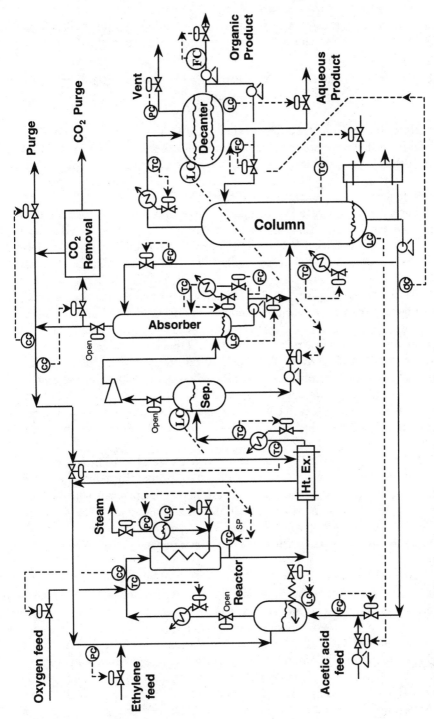

Figure 11.10 Vinyl acetate process control structure for "on-demand" product

2. The organic level in the decanter is controlled by manipulating the liquid flow from the upstream separator (the major feed stream to the column).

3. The level in the separator is controlled by changing the setpoint of the reactor temperature controller.

Note that in this structure levels are controlled in the opposite direction from flows. There is now a direct handle on production rate. When the organic product flowrate is changed, the plant responds by sequentially and gradually changing the flowrates back through the process. The last flowrates to change are the fresh feeds.

The production rate is ultimately determined by the conditions in the reactor. Thus, after a change in the product flow from the decanter there is a temporary difference between what is produced in the reactor and what is removed from the plant. This imbalance disappears after the reactor conditions are adjusted. For this scheme to work there must be enough internal buffer capacity to accommodate the reactor's temporary over-production or under-production of vinyl acetate.

The surge capacity of the separator had to be significantly increased from the values shown in Table 11.10. The liquid flowrate from the separator is 0.55 m^3/min. The working liquid volume in the separator at the base-case conditions is 8 m^3, which gives a holdup time of 14 min. Using the on-demand control scheme, this holdup time had to be increased to about 60 min to handle reasonable production rate changes (about 12%). The holdup time of the organic layer in the decanter was about 10 min.

In addition to providing sufficient surge capacity, care must be taken in tuning the separator level controller. The gain of this controller must be chosen such that the change in reactor temperature provides a reasonable change in the production of condensable material from the reactor. As with any cascade scheme, the primary (level) controller gain depends on the span of the secondary (temperature) controller transmitter.

Dynamic simulations of this on-demand control scheme verified that it works. However, a production rate change produces a dynamic response that is quite different from that shown in Fig. 11.4. For example the column is upset immediately whereas conditions in the reactor change slowly. Overall it appears that the internal time dynamics are slower for this on-demand scheme than when the reactor conditions are changed first. However, this control structure fulfills the control objective of providing immediate changes in the flowrate of the product stream.

11.6   Conclusion

The process considered in this chapter involved the production of vinyl acetate monomer. It features many unit operations, many components, nonideal phase equilibrium, unusual reaction kinetics, two recycle streams, and three fresh reactant makeup streams.

The plantwide process control design procedure was applied to this complex process; it yielded an effective base-level regulatory control system.

We showed through nonlinear dynamic simulations how the process reacts to various disturbances and changes in operating conditions. We have not shown any attempts to optimize process performance, to improve the process design, or to apply any advanced control techniques (model-based, nonlinear, feedforward, valve-position, etc.). These would be the natural next steps after the base-level regulatory control system had been developed to keep the process at a stable desired operating point.

11.7   References

Daniels, W. E. "Vinyl Ester Polymers" in *Encyclopedia of Polymer Science and Engineering*, 2d ed. New York: Wiley, Vol. 17, 393–425, (1989).

Fisher, W. R., Doherty, M. F., and Douglas, J. M. "The Interface Between Design and Control. 3. Selecting a Set of Controlled Variables," *Ind. Eng. Chem. Res.*, **27**, 611–615 (1988).

Guthrie, K. M. "Capital Cost Estimating," *Chem. Eng.*, **76**, 114–142, March 24, 1969.

Luyben, M. L., and Tyreus, B. D. "An Industrial Design/Control Study of the Vinyl Acetate Monomer Process," accepted for publication, *Comput. Chem. Eng.* (1998).

Samanos, B., Boutry, P., and Montarnal, R. "The Mechanism of Vinyl Acetate Formation by Gas-Phase Catalytic Ethylene Acetoxidation," *J. Catal.*, **23**, 19–30 (1971).

Smith, J. C. "The Solubility Diagrams for the Systems Ethylidene Diacetate–Acetic Acid–Water and Vinyl Acetate–Acetone–Water," *J. Phys. Chem.*, **46**, 229–232 (1942).

12

Conclusion

We have now completed our discussion of plantwide process control and our design procedure. We hope that you have found the chapters in this book informative and useful. We have attempted to provide a practical solution to a very important industrial problem. The concepts considered in this book should help you develop a workable control scheme for an entire chemical plant.

We have highlighted the features of complex integrated processes that force us to look at control from a plantwide perspective and not just combine the control loops of individual unit operations. These include material recycle, energy integration, and chemical component inventories. We also have examined the control of unit operations (reactors, heat exchangers, distillation columns, among others) within a plantwide context. Finally we have illustrated our plantwide control design procedure with four different industrial processes of varying degrees of complexity. Many more examples could be added to these four, but we feel these should be sufficient to demonstrate how the method can be applied and to prove its practical effectiveness. We have successfully applied this plantwide control design procedure to a number of other real industrial processes.

Our procedure may lack the mathematical sophistication and elegance preferred by many of today's academics in the field of process control. The emphasis for the last decade in academic research has focused primarily on the development of complex control algorithms and the exploration of their mathematical properties. Typically the examples considered in this academic research are very much simpler than a normal industrial chemical process.

One of our fundamental beliefs is that complex processes are best controlled by simple control systems. This view, however, directly contradicts the perspectives of many of today's academic researchers.

At the other end of the scale, our design procedure will not satisfy those who desire a single tool that can be broadly applied to any process by all engineers. Any such tool would necessarily be limited to a small set of well-defined processes. The complexity of a real process precludes us from being able to come to a single solution.

Another of our fundamental beliefs is that the control systems for complex processes can be designed only by engineers who understand the basic chemistry, physics, and economics of the process. We must be willing to think, not just compute. Process understanding is the key. It is vital always to keep the big picture in focus and not lose sight of the forest for the trees.

Our procedure is based on some very simple concepts: applying component balances on a plantwide basis; rejecting energy disturbances to the plant utility system; preventing large changes in recycle flowrates; finding variables that dominate reactor productivity and selecting one of these as a production rate manipulator to minimize the variability in product quality; and satisfying process, safety, and environmental constraints. It is completely heuristic in nature, not algorithmic. It is a synthesis tool for generating a control scheme for a given process flowsheet. It is not a synthesis tool for generating the "most controllable" process.

Active research programs in the area of plantwide control have begun to appear in several universities in recent years. We hope that more progress in this vital area will be made in the future. One of the most important unresolved problems is how to modify the process design to improve dynamic performance. These modifications could include alternative flowsheets, changes in design parameters, or operation at different conditions.

There has been some useful work in developing methods for comparing the economics of alternative designs, using both steady-state and dynamic considerations. The "capacity-based" economic approach of Elliott et al. (1997) provides a technique for quantitatively comparing the profitability of alternative processes.

In Chap. 1 we introduced the subject of plantwide process control by raising some questions concerning the control of the HDA process. From Sec. 1.2 these questions were

- How do we control the reactor temperature to prevent a runaway?

- How can we increase or decrease the production rate of benzene according to market conditions?

- How do we ensure the benzene product is sufficiently pure for us to sell?

- How do we know how much of the fresh hydrogen and toluene feed streams to add?
- How do we determine the flowrate of the gas purge stream?
- How can we minimize the raw material yield loss to diphenyl?
- How do we prevent overfilling any liquid vessels and overpressuring any units?
- How do we deal with units tied together with heat integration?
- How can we even test any control strategy that we might develop?

All of these questions, and many more, have been addressed in the chapters of this book.

Figures 12.1 through 12.11 give several flowsheets of typical chemical processes. You might find it instructive to apply the plantwide control design procedure to these processes. Very little process information is provided in these figures. You will have to make assumptions about a number of process conditions, flowrates, and control objectives. Keep a list of these as you work through the nine stages.

We hope that when you apply the design procedure presented in this book, you find it to be a useful guide to developing an effective control system for your plant. If you understand your process and if you keep

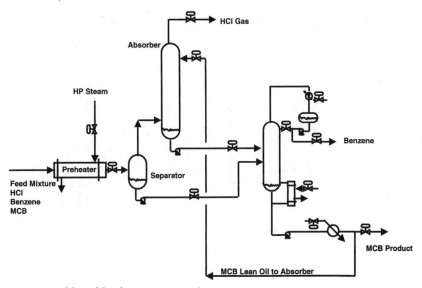

Figure 12.1 Monochlorobenzene separation.

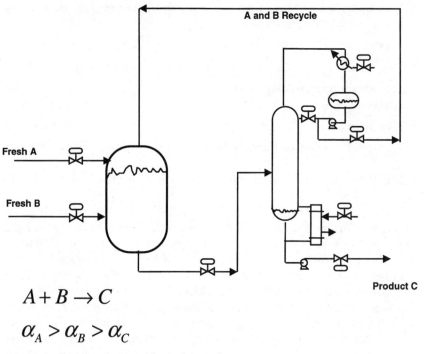

$$A + B \rightarrow C$$

$$\alpha_A > \alpha_B > \alpha_C$$

Figure 12.2 Ternary process with single recycle.

Fresh feeds are pure A and B, fed directly to reactor

$$A + B \Leftrightarrow C + D$$

$$\alpha_A > \alpha_C > \alpha_B > \alpha_D$$

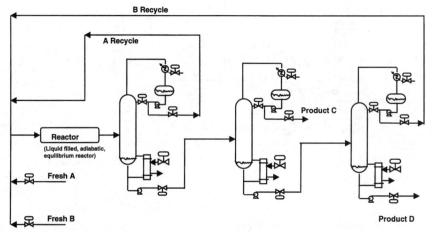

Figure 12.3 Process with two products and two recycles.

One fresh feed is mixture of A and C, fed to column

$$A + B \Leftrightarrow C + D$$

$$\alpha_A > \alpha_C > \alpha_B > \alpha_D$$

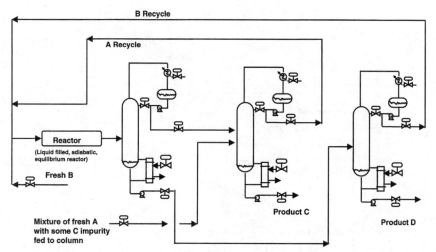

Figure 12.4 Process with two products and two recycles.

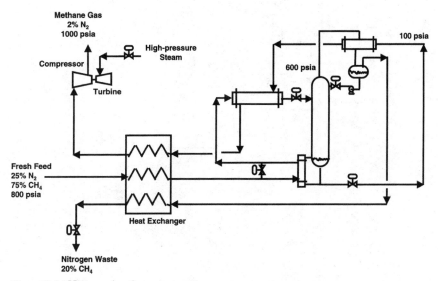

Figure 12.5 Nitrogen/methane separation.

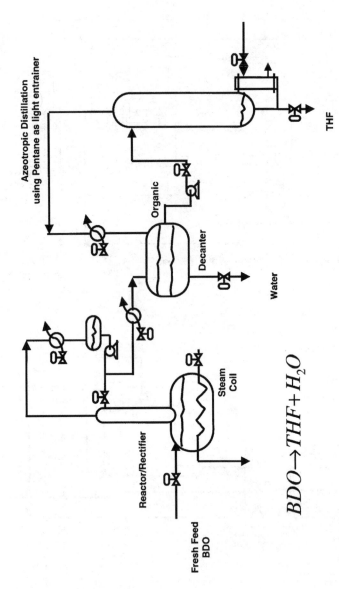

Figure 12.6 Tetrahydrofuran production.

Both M and D Produced

$$A + B \Leftrightarrow M + C$$

$$A + M \Leftrightarrow D + C$$

$$\alpha_A > \alpha_M > \alpha_D > \alpha_B > \alpha_C$$

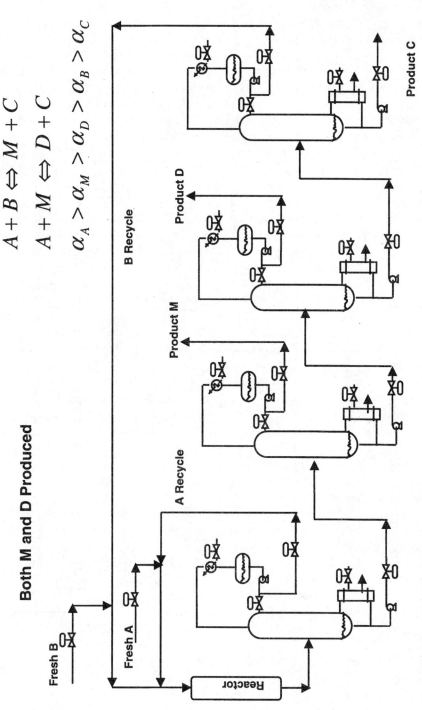

Figure 12.7 Process with consecutive reactions.

M Recycled to Extinction (Separate Recycles)

$$A + B \Leftrightarrow M + C$$

$$A + M \Leftrightarrow D + C$$

$$\alpha_A > \alpha_M > \alpha_D > \alpha_B > \alpha_C$$

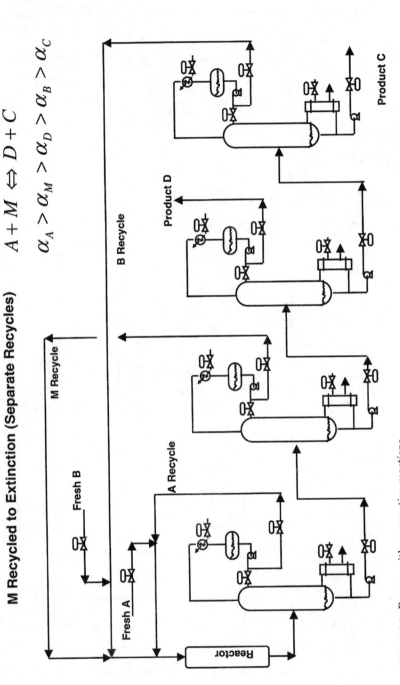

Figure 12.8   Process with consecutive reactions.

M Recycled to Extinction (Combined Recycle)

$$A + B \Leftrightarrow M + C$$

$$A + M \Leftrightarrow D + C$$

$$\alpha_A > \alpha_M > \alpha_D > \alpha_B > \alpha_C$$

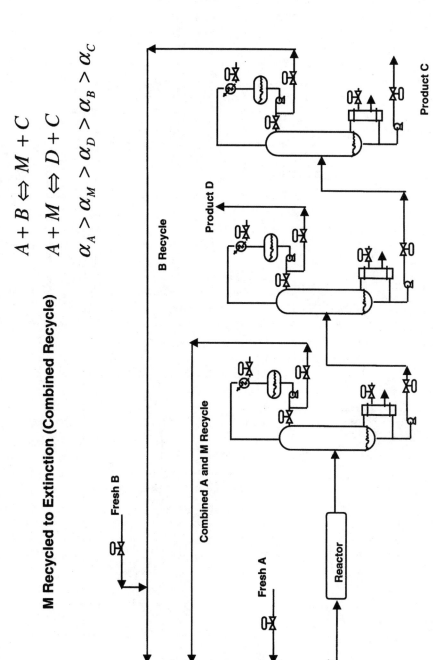

Figure 12.9   Process with consecutive reactions.

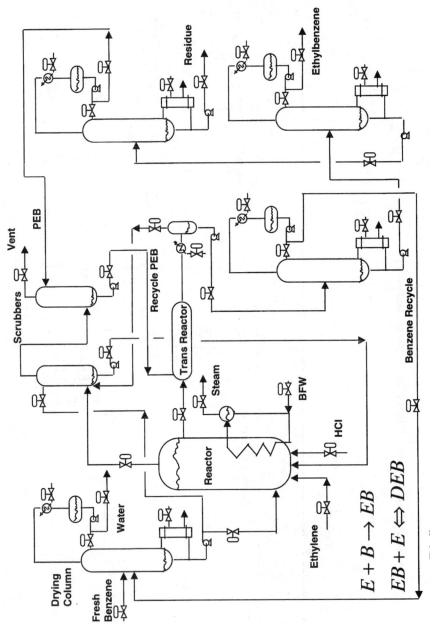

$E + B \rightarrow EB$

$EB + E \Leftrightarrow DEB$

Figure 12.10   Ethylbenzene process.

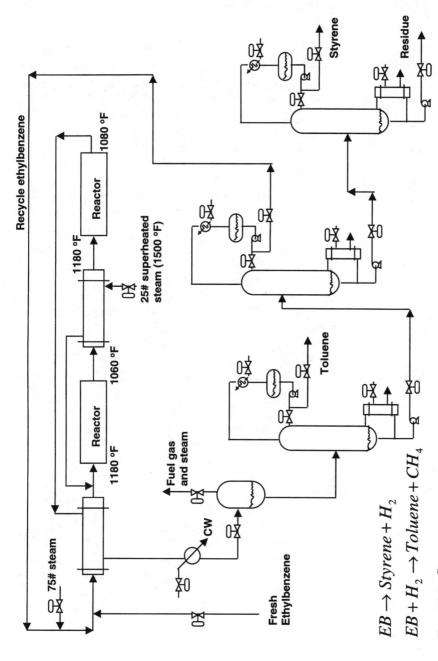

$$EB \rightarrow Styrene + H_2$$

$$EB + H_2 \rightarrow Toluene + CH_4$$

Figure 12.11   Styrene process.

the fundamental concepts of chemical engineering in mind, you will find a control system that provides stable base-level regulatory control of your process.

Reference

Elliott, T. R., Luyben, W. L., and Luyben, M. L. "Application of the Capacity-Based Economic Approach to an Industrial-Scale Process," *Ind. Eng. Chem. Res.*, **36,** 1727–1737 (1997).

Appendixes

A

Thermodynamics and Process Control

A.1 Introduction

It may seem surprising that we include an appendix on thermodynamics in a process control text. However, recent results have shown that there is a strong and important connection between the two fields. Some of these results have bearing on the energy management and control of chemical processes as referred to in the text. This appendix is therefore intended to help sort out the parts of thermodynamics that appear relevant to process control. For those interested in more details on thermodynamics we refer to one of the many excellent books on the subject (Denbigh, 1971; Callen, 1985; Carrington, 1994; Bejan, 1997).

A.2 Concepts

Thermodynamics at its highest level of abstraction involves only a few concepts and two laws. Yet it is able to produce a surprising number of useful results. We first examine the required concepts and then review the two laws in the next section.

The most fundamental concept in thermodynamics is that of a *system*. We may think of a system as a stream, a fluid holdup, a unit operation, or the entire plant. It is simply the part of the universe we carve out to study. The rest is the *environment*. A system is *closed* when no material enters or leaves, otherwise it is *open*. A closed system has a fixed mass and occupies a finite volume in space. Such a system is often called a *body*.

A second important concept is the system's *state*. The state is the system's physical condition of existence. It involves only quantities internal to the body or system. Quantities that contribute to determine the state must be available (in principle) by measurements made only on the body itself. For example, the kinetic energy of the entire body cannot be a function of state since the velocity that goes into the calculation of kinetic energy is defined in relation to other bodies.

The third concept to consider is how energy can flow between bodies. It can happen by only two mechanisms: by mechanical work performed on the body and by heating. Heating is energy flow driven by a temperature difference between two bodies.

A.3   The Laws of Thermodynamics

A.3.1   The first law

The first law of thermodynamics introduces a state function called *internal energy U*. Internal energy accounts for the energy forms in a body that depend on the body's state. It is important to remember that the internal energy is *not* the *total* energy of the body, which also includes the kinetic energy and the potential energy of the body as a whole, and these are defined in reference to other bodies. The internal energy is an *extensive* state variable, which means that its value is proportional to the body mass. The first law states that energy is conserved and is neither created nor destroyed by any transaction. Therefore flow of energy, either as incremental work done on the body, δw, or heat flow to the body, δq, will accumulate as an incremental increase in the internal energy. Mathematically we can describe this as follows:

$$dU = \delta w + \delta q \tag{A.1}$$

A.3.2   The second law

The second law of thermodynamics introduces another extensive state function, *entropy S*. The entropy function can be defined without any reference to the microscopic makeup of a system. However, such definitions make the entropy function abstract and hard to grasp since the word *entropy* is not part of familiar, everyday language (unlike energy, work, heat, system, etc.). It is therefore helpful to consider the meaning of entropy at a molecular level. Here it defines how the molecules are distributed (in a probabilistic sense) over all possible microscopic quantum states available to each molecule. Each quantum state represents a finite amount of energy, and the higher quantum states are more energetic. The entropy is high when, on average, a large portion of the available quantum states are occupied by molecules. We can force the molecules to occupy many quantum states by heating a body.

In particular, when heat is added to a body at absolute temperature T, the increase in entropy becomes

$$dS \geq \delta q / T \qquad (A.2)$$

The reason for the inequality sign is that entropy is not a conserved property. In fact, entropy is *produced* whenever spontaneous processes occur such as the irreversible flow of heat across a finite temperature differential. Only in the hypothetical case when heat is transferred *reversibly* to a body at constant temperature will the entropy increase equal the heat addition divided by the absolute temperature.

While heat and entropy are related according to Eq. (A.2) there is no corresponding relation between mechanical work and entropy. In fact, a reversible addition of mechanical energy to a body causes no change to the body's entropy. From a molecular viewpoint this can be understood by realizing that each microscopic quantum state is strengthened by the addition of mechanical energy. There is therefore no need for the molecules to redistribute towards high quantum states to account for the net increase in internal energy.

A.4 Heat, Work, and Exergy

A.4.1 Introduction

A typical chemical plant uses a large amount of energy to operate. Some of the energy is introduced in the form of heat from condensing steam and some is in the form of mechanical work (e.g., pumps and compressors driven by electricity). Although heat and work are two different energy forms, they cannot be interchanged arbitrarily. In fact, the famous scientists Carnot, Joule, and Clausius discovered in the nineteenth century that only a portion of the heat from a heat source can be turned into work while work can be completely turned into heat by friction. Work is therefore the "hard currency" of energy. When all the energy in a plant is "normalized" as work, it becomes clear that much of the useful work that goes in is actually lost due to inefficiencies in the process. An important design consideration is thus to minimize such inefficiencies. It is also of interest to know how effective the control system is in delivering useful work to the operation. We will therefore develop the appropriate tools to make a work analysis of a plant.

A.4.2 Fundamental property relation

Since the state of a system is a unique and repeatable condition, we wish to express all results in terms of state variables or changes in state variables. So far we have introduced three state variables, U, S, and T. Of these only temperature is practically measurable in a plant. But what about pressure and volume? They are also state variables that are practical to measure. To find relationships between the various state variables we apply the first and second laws to a closed system (a body). We first consider the reversible addition of heat to the body, $\delta q = TdS$. We also add work by reversibly compressing the body's volume V under constant pressure P. The work added to the body is $\delta w = -PdV$. Substitution into Eq. (A.1) gives

$$dU = TdS - PdV \qquad (A.3)$$

Equation (A.3) not only provides a differential relationship between state variables of interest but it also suggests that the internal energy is a *natural function* in two of the state variables, entropy and volume, $U(S,V)$. Indeed, it turns out that for a closed system we only need to specify the values of two state variables to fix uniquely all the other system states and properties. Furthermore, the function $U(S,V)$ is a *fundamental description* of the system because it allows us to calculate explicitly all other system properties. In practice we rarely have an analytical expression for the fundamental property relation. Even when we do, it is inconvenient to use because of its explicit dependency on entropy. Instead, *transformations* of the internal energy function can be derived that are natural in more convenient state variables. For example, the Gibbs free energy function, $G = U + PV - TS$, is a natural function in the measurable state variables temperature and pressure, $G(T,P)$. Since the Gibbs free energy function is merely a mathematical transformation of the internal energy function, the two energy functions are equally fundamental and informative of the system's state. The transformation is analogous to what is done in control theory when sets of linear differential equations are transformed into transfer functions. The transfer functions do not alter the information content but they are more convenient to manipulate analytically.

A.4.3   Maximum work from heat

We now investigate how much work we can derive from heat. Again, we only need the two laws of thermodynamics to do this. Our closed system will consist of an idealized heat engine (a Carnot engine) that can reversibly convert a fraction of heat into work. The system is "fueled" by a quantity of heat, q_i, transferred from a heat source operating at temperature T_i. The system produces w units of work while discharging the remaining heat q_0 to a sink operating at a lower temperature T_0. The maximum amount of work is produced when there is no accumulation of energy in the system. The first law gives the following relation between heat and work.

$$\Delta U = 0 = q_i - w - q_0 \tag{A.4}$$

The second law limits how much of the incoming heat can be turned into work. Specifically the second law states that *the entropy produced in the universe as a result of any process has to be greater than or equal to zero.* The entropy production inside our system is already zero (ideal heat engine assumed) but we must account for the entropy generation in the heat source and the heat sink. The second law requires

$$\Delta S_i + \Delta S_0 \geq 0 \tag{A.5}$$

Minimum entropy production occurs when the entropy change in the source cancels that of the sink and when the heat transfer between the system and the other bodies is done reversibly.

$$\Delta S_{0,min} = q_0/T_0 = -\Delta S_{i,min} = q_i/T_i \tag{A.6}$$

Elimination of q_0 between Eqs. (A.4) and (A.6) now gives the maximum amount of work that can be recovered from a unit of heat. Note that the result is expressed entirely in terms of the state variable temperature:

$$\frac{w_{max}}{q_i} = \frac{T_i - T_0}{T_i} \tag{A.7}$$

A.4.4 Maximum work from fluid system

We also need to know how much equivalent work is contained in a flowing stream or a holdup of fluid. Even here the two laws of thermodynamics are sufficient to derive the result. We choose our system to be closed but containing a fluid mixture. It may seem strange that we always select closed systems when we know that the majority of chemical plants are open and continuous. The reason is that a closed system provides all the relevant results and is easier to work with. It is generally trivial to extend the findings to open systems. In relation to this issue it will be practical to segregate the work done by the system into two categories. One portion contains shaft work and electrical energy and is symbolized by $-\delta w'$. The other part is mechanical work (*PV* work) that has to be extracted by a piston from expansion of the body at constant pressure (i.e., $-PdV$). For closed systems the *PV* work is impractical to extract and for open systems the *PV* work is involved in moving the streams in and out of the system.

By combining Eqs. (A.1) and (A.2), we obtain

$$dU - (\delta w' - PdV) \leq TdS \tag{A.8}$$

which can be rearranged to

$$-\delta w' \leq -dU - PdV + TdS \qquad (A.9)$$

The inequality accounts for irreversibilities that occur in practical systems. The theoretical, *maximum* attainable work must be extracted under reversible conditions for which equality holds:

$$-\delta w'_{max} = -dU - PdV + TdS \qquad (A.10)$$

To avoid explicit reference to the system's entropy we make use of the differentiated form of the Gibbs free energy function. We thus obtain

$$-\delta w'_{max} = -dG + VdP - SdT \qquad (A.11)$$

which at *constant temperature* and *pressure* simplifies to

$$-\delta w'_{max} = -dG \qquad (A.12)$$

or integrated

$$-w'_{max} = -\Delta G \qquad (A.13)$$

The change in Gibbs free energy is therefore the appropriate measure of the equivalent work in a fluid system operating under isobaric and isothermal conditions. Again, the result is expressed in terms of state variables (in this case G).

A.4.5 Exergy

Important as the Gibbs free energy function is in equilibrium thermodynamics, it is of somewhat limited use for our purposes since most practical processes don't have the same pressure and temperature in all streams and vessels. We therefore need an alternative way to express the work equivalent in a fluid system. Not too surprisingly, the first and second laws will again be our workhorses. Our system will contain a fluid mixture but is otherwise closed. The environment will be the conditions on the earth's surface, that is, the same as what we typically refer to as *the environment*.

We follow Denbigh (1956) by writing the first law relationship for an *integral* change in state. We assume that the process is heated from several heat sources operating at different temperatures T_i. The total heat load on the system is Σq_i energy units. We also assume that the environment is fixed and provides a constant pressure and temperature sink (P_0, T_0). The combined cooling load is q_0 energy units. The first law then states

$$\Delta U = -q_0 + \sum q_i - P_0 \Delta V + w \qquad (A.14)$$

where $\quad \Delta U =$ difference in internal energy between system's final and initial states

$\quad P_0\Delta V =$ work the system performs in displacing atmosphere at constant pressure

$\quad w =$ total of all other items of work done on system

Next, we account for the *total entropy change in the universe* σ as a result of operating the process.

$$\sigma \equiv \Delta S + \Delta S_0 + \sum \Delta S_i \geq 0 \qquad (A.15)$$

where $\quad \Delta S =$ change in entropy of fluids inside system

$\quad \Delta S_0 =$ increase in entropy experienced by environment

$\quad \Sigma \Delta S_i =$ sum of all entropy changes in heat sources

Since the environment receives heat at a constant temperature we can write $\Delta S_0 = q_0/T_0$. The cooling load q_0 can thus be eliminated between Eqs. (A.14) and (A.15) to give

$$\Delta S + \frac{1}{T_0}\left(-\Delta U + \sum q_i - P_0\Delta V + w\right) + \sum \Delta S_i = \sigma \qquad (A.16)$$

which can be rearranged to

$$w + \sum \left(q_i + T_0\Delta S_i\right)$$

$$= T_0\sigma + \Delta U + P_0\Delta V - T_0\Delta S = T_0\sigma + \Delta B \qquad (A.17)$$

The left hand side of Eq. (A.17) represents the total work performed on the process. This can be visualized by replacing ΔS_i with $-q_i/T_i$ such that $q_i + T_0\Delta S_i = q_i(T_i - T_0)/T_i$, which equals the net work from a Carnot engine, as we showed in Eq. (A.7). The right hand side of Eq. (A.17) designates how the work is used by the process. Part of it goes into changing the state of the fluid within the process as represented by ΔB. The rest is lost due to irreversibilities resulting from carrying out the process. The maximum work available in a fluid system is therefore obtained when the whole process is reversible ($\sigma = 0$). We then have

$$-w_{\max} = -\Delta B \qquad (A.18)$$

The new state function *exergy*, B, giving the work equivalent of a fluid system operating between any set of conditions, is defined as

$$B = U + P_0V - T_0S - (U_0 + P_0V_0 - T_0S_0) \qquad (A.19)$$

where U_0, V_0, S_0 are the internal energy, volume, and entropy of the components in the environment with which our system would come to equilibrium if all spontaneous processes were carried out. In words, exergy is defined as "the amount of work obtainable when some matter is brought to a state of thermodynamic equilibrium with the common components of the natural surroundings by means of reversible processes, involving interactions only with the above-mentioned components of nature" (Szargut et al., 1988).

A.5   Thermodynamics and Process Design

A chemical process converts raw materials into products. The exergy content of the raw materials is usually greater than the exergy content of the products. Equation (A.18) then indicates that we should be able to perform the process and be a net producer of work. Reality tells us differently. Most chemical plants require a substantial energy input, making them extremely inefficient from a thermodynamic standpoint. The reason is that most practical and economical unit operations have substantial inefficiencies as part of their operation resulting in exergy *destruction* (or entropy production) (Bejan, 1997). Entropy is produced inside a process for the following reasons:

1. Pressure drop and friction in pipes and vessels

2. Heat transfer across finite temperature differences

3. Mixing of streams of different temperatures

4. Mixing of streams of different compositions

5. Irreversible reactions

In addition, exergy is destroyed whenever heat is rejected to the environment (energy *dissipation*).

From the viewpoint of steady-state operating cost, it is desirable to improve the thermodynamic efficiency of a chemical process. This translates to reduced energy costs. However, it often takes capital to implement the efficiency improvements. Let's consider a chemical reactor, for example. If the reaction is exothermic and we cool it with cooling water, we destroy exergy by dissipating energy to the environment. The only way we could retain most of the exergy is if we carried out the reaction electrochemically. Such reactor designs would be complicated and expensive at best and most likely infeasible with today's technology. Instead, we could use a conventional reactor and capture the heat in a waste heat boiler. The steam produced can then be used for heating or to drive a turbine to generate work. The waste heat boiler reduces the plant's net energy consumption but adds to the overall investment cost.

Similar considerations apply to distillation columns. Entropy is generated on the trays due to mixing and heat transfer between dissimilar streams. We can reduce the entropy production by adding more trays to the column. With more trays there is less of a gradient across each tray and the entropy production goes down. A column with many trays requires less reflux and therefore less heat to the reboiler. The energy consumption is reduced only at the expense of added capital investment for the taller column.

A thermally efficient process does not always have to be the most expensive from an investment standpoint. Consider heat-integrated plants as an example. Here heat-producing operations are integrated with heat-consuming units. The objective is to avoid dissipating high-temperature heat to the environment when the energy can be used elsewhere in the plant. Such heat recovery schemes not only save on the plant's net heating and cooling load, but they may also reduce the capital expense by eliminating two utility exchangers for each process-to-process exchanger installed.

A.6 Thermodynamics and Process Control

From a steady-state energy consumption viewpoint, we always strive to make a process as efficient as possible. The only reason we stop at a certain level is the trade-off with capital costs as indicated above. But what about operability and control? Is a thermodynamically efficient process also easy to control? If so we may be able to justify an even higher level of efficiency than dictated by steady-state economics alone. On the other hand, if thermodynamic efficiency makes the process harder to control, we have another trade-off to consider during process design.

To investigate this issue we need to consider process dynamics. This is usually done by writing the differential equations describing the accumulation of material and internal energy inside a confined volume (Luyben, 1990). Each compartment in a process will then have $N + 1$ *dynamic state variables*, where N is the number of components in the system. A large number of dynamic state variables are usually required to describe the behavior of a real process. For example, assume that we are describing the dynamics of a process containing 9 components. This means that a plug-flow reactor with 20 compartments has 200 state variables and a distillation column with 40 trays has 420 state variables when we include the reboiler and condenser dynamics. A small plant with one reactor and a couple of columns may then require over 1000 dynamic state variables to capture its dynamic behavior. How can we possibly draw any general conclusions from large systems like that?

The most common approach to this problem is to perform numerical simulations on the nonlinear process model. This provides quantitative answers to specific questions regarding the process at hand but does not necessarily answer the more general question we are asking regarding thermodynamic efficiency and controllability. Instead we will explore what we can learn directly from thermodynamics.

Since thermodynamic efficiency is linked to the destruction of exergy and controllability is an issue of process dynamics, we start our investigation by writing a dynamic exergy balance. This can easily be done by differentiation of Eq. (A.17) and adding exergy flow terms to allow for representations of open systems (Bejan, 1997).

$$\frac{dB(\mathbf{x}(t))}{dt} = +[B_f(t) + \dot{w} + \sum \dot{q}_i(T_i - T_0)/T_i]_{in}$$

$$-[B_f(t) + \dot{w} + \sum \dot{q}_i(T_i - T_0)/T_i]_{out} - T_0\dot{\sigma}(\mathbf{x}(t)) \quad (A.20)$$

$$= \dot{\phi} - T_0\dot{\sigma}(\mathbf{x}(t))$$

where $dB(\mathbf{x}(t))/dt$ = rate of exergy accumulation within system
$B_f(t)$ = exergy flow carried by fluid streams entering
        or leaving system
$\dot{w}$ = mechanical power delivered to or derived
      from system
$\dot{q}_i$ = heat input (or removal) rate from source i
$\dot{\sigma}(\mathbf{x}(t))$ = total entropy production rate as a result
      of running process
$\dot{\phi}$ = net input of exergy to system

By $B(\mathbf{x}(t))$ and $\dot{\sigma}(\mathbf{x}(t))$ we have indicated that both the exergy content and the entropy production rate are functions of the state variables $\mathbf{x}(t)$ in the system.

Before we can demonstrate the connection between process control and Eq. (A.20), we need to introduce the concept of *Lyapunov functions* (Schultz and Melsa, 1967). Lyapunov functions were originally designed to study the stability of dynamic systems. A Lyapunov function is a positive scalar that depends upon the system's state. In addition, a Lyapunov function has a negative time derivative indicative of the system's "drive" toward its stable operating point where the Lyapunov function becomes zero. Mathematically we can describe these conditions as

$$V(\mathbf{x}(t)) > 0, \mathbf{x}(t) \neq \mathbf{x}^*$$

$$\frac{dV(\mathbf{x}(t))}{dt} < 0, \mathbf{x}(t) \neq \mathbf{x}^*$$

$$V(\mathbf{x}^*) = 0$$

where $V(\mathbf{x}(t))$ = Lyapunov function and $\mathbf{x}^*$ = state vector at reference operating point.

If it is possible to find a Lyapunov function for a dynamic system operating around the reference state $\mathbf{x}^*$, it follows that $\mathbf{x}^*$ is a stable state that is approached asymptotically. We can demonstrate this technique by verifying the stability of the reference state corresponding to the earth's environment. The exergy function is a suitable Lyapunov function for the following three reasons. First, exergy is a state function that depends on all other thermodynamic state variables in the system. Second, by definition, the exergy is zero at the environment state and nonzero for all other states [see Eq. (A.19), $B(\mathbf{x}^*) = 0$ and $B(\mathbf{x}(t)) > 0$, $\mathbf{x}(t) \neq \mathbf{x}^*$]. Finally, Eq. (A.20) shows that an unforced ($\phi = 0$) and unconstrained system has a strictly negative time derivative since the entropy production is positive for all spontaneous processes in nature leading toward equilibrium.

$$\frac{dB(\mathbf{x}(t))}{dt} = -T_0\dot{\sigma}(\mathbf{x}(t)) < 0, \, \mathbf{x}(t) \neq \mathbf{x}^* \qquad (A.21)$$

These conditions are sufficient to demonstrate the stability of the environment state (which we already know is an equilibrium state).

We now turn our attention to actual process systems operating away from equilibrium conditions. Such systems are forced ($\phi \neq 0$), and while stationary are not at equilibrium. How can we investigate their properties through thermodynamics? The answer lies in an important result in systems theory due to Ydstie and Alonso (1997), who showed that exergy is a valid *storage* function for process systems and that such systems are *dissipative*.

A storage function is a nonnegative entity that can accumulate in the system and depends upon the system's state. Valid storage functions must possess certain mathematical properties as explained by Willems (1974). Energy is the most commonly used storage function for mechanical and electrical systems. For example, in a mechanical system the sum of kinetic and potential energy is a storage function. Similarly, in electrical systems, the power imposed from the product of voltage and current accumulated over time is a legitimate storage function.

A stored quantity can be produced, conserved, or dissipated. For example, the sum of kinetic and potential energy is conserved in a mechanical system without friction. However, the presence of friction will dissipate some of the useful mechanical energy. Similarly, in a network of capacitors and resistors, only part of the supplied energy is stored in the capacitors while the rest is dissipated as heat in the resistors.

From a control standpoint it is desirable to find dissipative storage functions where the net *supply* of the stored quantity is a function of the system's measured outputs and manipulated inputs. Such systems are said to be *passive*. The notion of passivity comes from the behavior of electronic circuits consisting of interconnected passive components (e.g., capacitors and resistors). Circuits containing active components (e.g., transistors and operational amplifiers) are consequently not passive. Passive systems can be proven to be asymptotically stable because the *dissipative storage function is a Lyapunov function*. Therefore, it is of great interest to investigate how one can make an arbitrary dynamic system passive. This involves three steps. First, a suitable storage function must be identified. Second, measurements on the system's inputs and outputs must indicate the net supply of the stored quantity as well as the amount of storage within the system. Finally, it must be possible to adjust the manipulated variables such that the net supply can be held constant. Examples of this methodology are given by Ydstie and Viswanath (1994).

The control implication of exergy being a dissipative storage function for process systems is that, if we could find ways of measuring the exergy content B and the net exergy supply ϕ to a process and then find ways of manipulating the inputs to deliver this supply, we would have a passive system. The net exergy input could be set at an arbitrary value and the process system would find a unique, stable steady state because the exergy content is a Lyapunov function. Furthermore, at each steady state the system's exergy content and exergy destruction rate (entropy production) would be at a global minimum and all dynamic state trajectories would approach this minimum asymptotically.

We may now ask if there is a way of telling how quickly this minimum is reached following a disturbance? Because the steady state is at an exergy minimum and the entropy production rate is also at its lowest value, both these entities would have to increase when the system is perturbed away from the steady state. If we designate the increase in system exergy by $\Delta B(\mathbf{x}(t))$ and the increase in the exergy destruction rate by $T_0 \Delta \dot{\sigma}(\mathbf{x}(t))$, we may define a system settling time or response time τ by considering the ratio between the system's exergy change and the change in rate of exergy destruction *for all possible state trajectories close to the steady state*:

$$\tau = \max_{\mathbf{x}} \frac{\Delta B(\mathbf{x}(t))}{T_0 \Delta \dot{\sigma}(\mathbf{x}(t))} < \infty \qquad (A.22)$$

Equation (A.22) indicates that the system response is bounded by the time constant τ, which measures how quickly a small increase in the system's exergy storage would dissipate due to a corresponding increase in the rate of entropy production. A small time constant means a fast system response and an opportunity for tight control. This occurs when the entropy production rate increases substantially for small increases in the exergy storage.

We thus arrive at an interesting conclusion regarding thermodynamics and process control. It is not the steady state irreversibility (inefficiency) that matters for control but the ability to alter the rate of total entropy production in response to the system's departure from steady state. We have previously indicated qualitatively how entropy is produced. To see how the rate of entropy production changes with the system's state, we need to perform a quantitative analysis. This requires a brief introduction to the subject of *nonequilibrium thermodynamics* (Callen, 1985; Haase, 1990).

A.7 Nonequilibrium Thermodynamics

Earlier we stated that the fundamental property relation for a closed, equilibrium system has the form $U(S,V)$. When we allow the system to be open we must include the effects of varying mass and composition on the internal energy. It turns out that the extended property relation for an open, equilibrium system is $U(S,V,N_1,N_2,\ldots,N_n)$ where N_i is the mole number of component i. We can differentiate this function and compare it to Eq. (A.3) to obtain

$$dU = TdS - PdV + \sum_{i=1}^{n} \mu_i dN_i \qquad (A.23)$$

where the chemical potential is defined as

$$\mu_i \equiv \left(\frac{\partial U}{\partial N_i}\right)_{S,V,N_1,N_2,N_{i-1},N_{i+1},N_n} \qquad (A.24)$$

Equation (A.23) can be rearranged to make it explicit in entropy.

$$dS = \frac{1}{T}dU + \frac{P}{T}dV - \sum_{i=1}^{n} \frac{\mu_i}{T} dN_i \qquad (A.25)$$

Equations (A.23) and (A.25) pertain to equilibrium conditions of homogeneous systems. Such systems have constant properties over space and time and there is no entropy production. We shall now be interested in systems, away from equilibrium where properties vary as functions of location as well as time. To apply the results of thermodynamics to nonequilibrium systems, the principle of local (microscopic) equilibrium is invoked. For that reason it is useful to work with the thermodynamic variables on a unit volume basis. Equation (A.25) then becomes

$$ds = \frac{1}{T} du - \sum_{i=1}^{n} \frac{\mu_i}{T} dc_i \tag{A.26}$$

where ds = total change in local entropy density
du = total change in local internal energy density
dc_i = total change in local molar concentration
of component i

Equation (A.26) is used to find an expression for the time variation of the local entropy density (Haase, 1990).

$$\frac{\partial s}{\partial t} = \left(\frac{1}{T}\right) \frac{\partial u}{\partial t} - \sum_{i=1}^{n} \left(\frac{\mu_i}{T}\right) \frac{\partial c_i}{\partial t} \tag{A.27}$$

In addition, it is possible to derive an expression for the time variation of the local entropy density by writing a dynamic entropy balance:

$$\underbrace{\frac{\partial\left(\int_V s\,dV\right)}{\partial t}}_{\text{accumulation}} = - \underbrace{\int_A \mathbf{J}_S \cdot \hat{\mathbf{n}}\,dA}_{\text{net outflow}} + \underbrace{\int_V \dot{\sigma}_V dV}_{\text{production}} \tag{A.28}$$

where $\mathbf{J}_S$ = space vector designating net entropy flow per unit area leaving volume element dV
$\hat{\mathbf{n}}$ = unit vector perpendicular to surface dA surrounding volume element
$\dot{\sigma}_V$ = *local* entropy production density as a result of running process

Equation (A.28) can be simplified to

$$\frac{\partial s}{\partial t} = - \nabla \cdot \mathbf{J}_S + \dot{\sigma}_V \tag{A.29}$$

where the following two mathematical identities have been used:

$$\frac{\partial\left(\int_V s\,dV\right)}{\partial t} = \int_V \left(\frac{\partial s}{\partial t}\right) dV$$

$$\int_A \mathbf{J}_S \cdot \hat{\mathbf{n}}\,dA = \int_V (\nabla \cdot \mathbf{J}_S)\,dV$$

The second identity follows from the *divergence theorem* where ∇ symbolizes the operator

$$\left(\mathbf{i} \frac{\partial}{\partial x} + \mathbf{j} \frac{\partial}{\partial y} + \mathbf{k} \frac{\partial}{\partial z} \right)$$

Equation (A.26) suggests that we can relate the entropy flow vector $\mathbf{J}_S$ to the energy flow vector $\mathbf{J}_e$ and the component flow vectors $\mathbf{J}_i$ according to

$$\mathbf{J}_S = \frac{1}{T} \mathbf{J}_e - \sum_{i=1}^{n} \frac{\mu_i}{T} \mathbf{J}_i \qquad (A.30)$$

By setting Eqs. (A.27) and (A.29) equal, we obtain an expression for the local entropy production term:

$$\dot{\sigma}_V = \left(\frac{1}{T} \right) \frac{\partial u}{\partial t} - \sum_{i=1}^{n} \left(\frac{\mu_i}{T} \right) \frac{\partial c_i}{\partial t} + \nabla \cdot \left(\frac{1}{T} \mathbf{J}_e - \sum_{i=1}^{n} \frac{\mu_i}{T} \mathbf{J}_i \right) \qquad (A.31)$$

The rate of change in the local internal energy density, $\partial u / \partial t$, and the component concentrations, $\partial c_i / \partial t$, are obtained from an energy balance and n component balances similar to Eq. (A.29):

$$\frac{\partial u}{\partial t} = -\nabla \cdot \mathbf{J}_e + q_u$$

$$\frac{\partial c_i}{\partial t} = -\nabla \cdot \mathbf{J}_i + \sum_{j=1}^{n_r} v_{ij} r_j \qquad (A.32)$$

where q_u = generation of internal energy (e.g., frictional heat) due to degradation of potential and kinetic energies in flowing fluid

$\sum_{j=1}^{n_r} v_{ij} r_j$ = production of component i due to n_r chemical reactions in system

The final expression for the local entropy production density now becomes

$$\dot{\sigma}_V = \mathbf{grad} \left(\frac{1}{T} \right) \cdot \mathbf{J}_e -$$

$$\sum_{i=1}^{n} \mathbf{grad} \left(\frac{\mu_i}{T} \right) \cdot \mathbf{J}_i + \frac{1}{T} \sum_{j=1}^{n_r} A_j r_j + \frac{1}{T} q_u \geq 0 \qquad (A.33)$$

where $\mathbf{grad}\,(x)$ = vector measuring gradient of scalar quantity x

$$A_j = \textit{affinity} \text{ of reaction } j \text{ and equals } \left(- \sum_{i=1}^{n} v_{ij}\mu_i \right)$$

The total entropy production as needed in Equation (A.22) is related to the entropy density as follows:

$$\dot{\sigma} \equiv \int_V \dot{\sigma}_V dV$$

We now examine some of the important features of Eq. (A.33).

A.7.1   Forces and fluxes

The first feature concerns the structure of the terms in Eq. (A.33). Each term can be viewed as the product between a *generalized (driving) force* X_k and a *generalized flux* J_k. The first term in Eq. (A.33) has the temperature gradient as a force and heat transfer rate as a flux. The second term has a composition gradient and a mass transfer flux. The third term has affinity as a force (indicative of the distance away from chemical equilibrium) and reaction rate as the flux. The fourth term is already a composite related to pressure drop and fluid flow. Equation (A.33) can therefore be written compactly as

$$\dot{\sigma}_V = \sum_k X_k J_k \geq 0 \tag{A.34}$$

Forces and fluxes are not independent. We know from observations that a flux depends upon the magnitude of the corresponding force. For example, in the linear regime the flux is directly proportional to the driving force (e.g., Fick's and Fourier's laws of mass and heat transfer). This means that the entropy production rate increases nonlinearly with imposed forces (gradients). However, the presence of a flux tends to reduce the gradient. For example, consider dissolving a solid in a fixed amount of pure solvent. Initially, the concentration gradient is large and the solid goes into solution quickly. Later, as the part of the solid dissolves, its concentration in the solvent increases and the flux decreases. The flux eventually vanishes when the solvent is saturated (zero gradient).

A.7.2   Coupling

The second law of thermodynamics requires that the entropy produced from running a process must be positive. For example, when we impose a temperature gradient across a body, heat flows from hot to cold such that the product of the gradient and the flux is positive.

$$\mathbf{grad}\left(\frac{1}{T}\right) \cdot \mathbf{J}_e = -\frac{1}{T^2}\,\mathbf{grad}\,(T) \cdot \mathbf{J}_e \geq 0$$

The same can be said about material transport, chemical reactions, and friction due to pressure drop when these phenomena are studied one at a time. However, in unit operations we usually have several phenomena taking place simultaneously. The second law requires that only the *total* entropy production is positive; not that each term in Eq. (A.33) has to be positive. In fact, we often have situations where one term is negative at the expense of an even larger positive term. In continuous distillation, for example, we can purify a feed stream against the components' concentration gradients because a large heat flux travels from the hot reboiler to the cold condenser. Contrast this with an adiabatic tubular reactor with an exothermic reaction, where heat flows *from* the *cold* inlet *to* the *hot* exit because the reaction affinities and rates dominate Eq. (A.33). This interplay between terms in the overall entropy production equation is called *indirect coupling* and can affect the characteristics and controllability of a unit operation. For example, indirect coupling can interfere with the control system for complex reactors carrying out complex reactions. Since the reaction rates are influenced by temperature as well as composition, it is important to control these variables. However, when the reactor has composition, temperature, flow, and pressure gradients, all the thermodynamic state variables interact through coupling to produce unpredictable and uncontrollable results. The problem can best be solved by design simplifications either in the chemistry or in the reactor itself.

A.7.3 Controllability Implications

We use the system response time τ as a measure of controllability. We found that a short response time could be achieved for large changes in the exergy destruction rate $T_0\Delta\dot{\sigma}$, relative to the change in the stored exergy ΔB. We now examine how these changes are related to the system's state.

Equation (A.19) gives the change in exergy in terms of other thermodynamic variables:

$$\Delta B = \Delta U + P_0\Delta V - T_0\Delta S$$

This implies that there is an exergy increase whenever the internal energy increases (even from heat, since $q \approx T\Delta S > T_0\Delta S$). We must therefore conclude that the exergy storage will go up with increases in average temperature, pressure, and concentrations of the chemical components.

We next turn to the change in exergy destruction rate due to departures from steady state. Equation (A.33) gives

$$T_0 \Delta \dot{\sigma} = T_0 \int_V (\Delta \dot{\sigma}_V)\, dV$$

$$\Delta \dot{\sigma}_V = \Delta \left(\mathbf{grad}\left(\frac{1}{T} \right) \cdot \mathbf{J}_e \right) - \sum_{i=1}^{n} \Delta \left(\mathbf{grad}\left(\frac{\mu_i}{T} \right) \cdot \mathbf{J}_i \right)$$

$$+ \frac{1}{T} \sum_{j=1}^{n_r} \Delta\,(A_j r_j) + \frac{1}{T} \Delta q_u \tag{A.35}$$

which can be written in a compact and simplified form omitting higher-order terms by the use of Eq. (A.34):

$$T_0 \Delta \dot{\sigma} \approx T_0 \int_V dV \sum_k \Delta X_k \overline{J}_k + T_0 \int_V dV \sum_k \overline{X}_k \Delta J_k \tag{A.36}$$

Equation (A.36) shows that the exergy destruction rate increases with changes in the gradients as well as the fluxes. There are interesting control implications from this relationship. We outline two of them here. Others are mentioned in the main text in connection with the control strategies for specific unit operations.

1. *The steady-state gradient should be designed to be small whenever the gradient is manipulated for control.* This follows from the dependency of fluxes on forces:

$$\overline{J}_k \approx \alpha \overline{X}_k$$

$$T_0 \Delta \dot{\sigma} \approx 2 T_0 \int_V dV \sum_k \alpha \overline{X}_k \Delta X_k \tag{A.37}$$

When the gradient $\overline{X}_k$ is small, some other factor α must be designed to be large to obtain the required flux $\overline{J}_k$. Take an exothermic reactor, for example. A small temperature gradient between the reactor content and the cooling medium requires a large heat transfer area to satisfy the cooling requirement (the flux). Now, a small increase in the reactor temperature results in a large relative change in the temperature gradient. The reactor temperature increase causes a modest change in the stored exergy, whereas the gradient makes the entropy production rate increase significantly. These two effects promote controllability. Similarly for an isothermal, irreversible reactor under composition control, a small affinity (nearly complete conversion) requires a large reactor. Small increases in the concentrations of the reacting components will then cause significant relative increases in the affinity, thus enhancing the entropy production rate and improving control.

2. *A large steady-state gradient is desirable when the flux is manipulated for control.* This follows directly from Eq. (A.36) when the gradients remain constant.

$$T_0 \Delta \dot{\sigma} \approx T_0 \int_V dV \sum_k \overline{X}_k \Delta J_k \qquad (A.38)$$

Mixing processes and distillation columns benefit from this principle. For example, tray temperature control of a distillation column with a large temperature gradient is much quicker than the control of a column separating close-boiling isomers. Similarly, blending a small stream of a pure component into a large mixed stream is more responsive than the mixing of two similar streams. Also, adjusting the flow through a control valve is effective when there is a large pressure drop across the valve.

A.8   Conclusion

We have used the fundamental principles of thermodynamics to demonstrate how the exergy function is important in evaluating the steady-state efficiency of a process. Specifically, exergy is the "hard currency" of energy and is therefore the right measure for differentiating between various energy sources with the same energy content. An exergy analysis of a chemical process can give important insights into where energy is "wasted" and suggest measures for efficiency improvements.

We have also shown that, in addition to being a measure of thermodynamic efficiency, the exergy function plays a role in evaluating the controllability of a plant. In this context it appears important to have effective handles on the exergy destruction rate (total entropy production) to respond to changes in the dynamic state of the process. We do not necessarily imply that a steady-state efficient process has to be less controllable than an inefficient one, but we claim that a controllable process must have the means to change its exergy destruction rate radically from the nominal value. These principles are demonstrated throughout the book.

A.9   References

Bejan, A. *Advanced Engineering Thermodynamics*, 2nd ed., New York: Wiley (1997).
Callen, H. B. *Thermodynamics and an Introduction to Thermostatistics*, 2nd ed., New York: Wiley (1985).
Carrington, G. *Basic Thermodynamics*, New York: Oxford University Press (1994).

Denbigh, K. G. "The Second-law Efficiency of Chemical Processes," *Chem. Engng. Sci.*, **6,** 1–9 (1956).

Denbigh, K. G. *The Principles of Chemical Equilibrium,* 3d ed., New York: Cambridge University Press (1971).

Haase, R. *Thermodynamics of Irreversible Processes,* New York: Dover Publications (1990).

Luyben, W. L. *Process Modeling, Simulation and Control for Chemical Engineers,* 2d ed., New York: McGraw-Hill (1990).

Schultz, D. G., and Melsa, J. L. *State Functions and Linear Control Systems,* New York: McGraw-Hill (1967).

Szargut, J., Morris, D. R., and Steward, F. R. *Exergy Analysis of Thermal, Chemical, and Metallurgical Processes,* New York: Hemisphere (1988).

Ydstie, B. E., and Alonso, A. A. "Process Systems and Passivity via the Clausius-Plank Inequality," *Systems & Control Letters,* **30,** 253–564 (1997).

Ydstie, B. E., and Viswanath, K. P. "From Thermodynamics to a Macroscopic Theory for Process Control," paper presented at AIChE Meeting, November 1994.

Willems, J. C. "Dissipative Dynamical Systems, Part I: General Theory," *Arch. Rational Mech. Anal.,* **45,** 321–350 (1974).

B

Nonlinear Plantwide Dynamic Simulations

Nonlinear dynamic simulations of plantwide systems can be performed using a variety of software packages and computer platforms. We can write our own program to integrate numerically the differential equations describing the system. We also can use one of the commercial modeling programs that are now available.

But even if we have a program and after we have gone through the design procedure outlined in this book, the mechanics and techniques for performing a plantwide control study are not necessarily obvious. Where and how do we start? We attempt in this appendix to provide a little guidance for this not-so-simple process. Our suggested line of attack is as follows:

1. Start with one isolated unit operation. Get its control system installed, tuned, and tested for closed-loop stability and robustness. Then move on to the next unit and repeat. Build up the entire plant one unit at a time until you get all recycles connected.

2. Initially use proportional-only controllers in all loops except flow controllers, where the normal tight tuning can be used ($K_c = 0.5$ and $\tau_I = 0.3$ minutes). Set the gains in all level controllers (except reactors) equal to 2. Adjust the temperature, pressure, and composition controller gains by trial and error to see if you can line out the system with the proposed control structure. If P-only control cannot be made to work, PI will not work either. When stable operation is achieved, add a little reset action to each PI controller (one at a time) to pull the process into the setpoint values.

3. Perform a relay-feedback test on each temperature, pressure, and

composition loop one at a time to determine the ultimate gains and frequencies. Remember to place reasonable lags and deadtimes in these loops in your simulation program. Temperature measurement lags are 0.3 to 0.5 minutes in liquid streams and 0.6 to 1.5 minutes in gas streams. The relay-feedback test does not work unless the loop has a phase angle less than $-180°$, so the process transfer function must be at least third-order or contain deadtime.

4. Use PI controllers in most temperature, pressure, and composition loops. Derivative action can help improve dynamics in some noise-free situations, but if you cannot get good control using only PI in a simulation environment, it is doubtful that the control of a real plant will be good.

5. Ziegler-Nichols settings should be used only if tight control is desired and large swings in the manipulated variable can be tolerated. Reactor temperature control is a typical example:

$$K_c = K_u/2.2$$

$$\tau_I = P_u/1.2$$

6. Use Tyreus-Luyben settings for most loops, particularly distillation columns where large swings in vapor boilup or reflux are undesirable. These settings are much more conservative (robust) than Ziegler-Nichols:

$$K_c = K_u/3.2$$

$$\tau_I = 2.2P_u$$

7. Once the entire plant is running, use different disturbances with different magnitudes and in different directions to test the control structure. If large changes in manipulated variables are required to get to the new steady state, the control structure will perform poorly.

INDEX

ABOUT THE AUTHORS

WILLIAM L. LUYBEN, Ph.D., is a professor of chemical engineering at Lehigh University and the author or coauthor of six textbooks, including *Essential Process Control* and *Process Modeling,* and over 150 technical papers.

BJÖRN D. TYRÉUS, Ph.D., is a research fellow in Du Pont's Central Research and Development Department in Wilmington, Delaware.

MICHAEL L. LUYBEN, Ph.D., is a senior research engineer at Du Pont's Central Research and Development Department. They are both authors of a number of technical papers on the interaction of process design and process control. Michael and William Luyben are coauthors of *Essentials of Process Control,* published by McGraw-Hill in 1997.